Towards Functional Safety in Drive-by-Wire Vehicles

Peter Johannes Bergmiller

Towards Functional Safety in Drive-by-Wire Vehicles

This dissertation was submitted to and accepted
at Technische Universität Braunschweig,
Braunschweig, Germany

Springer

Peter Johannes Bergmiller
Institut für Regelungstechnik
TU Braunschweig
Braunschweig, Bayern
Germany

ISBN 978-3-319-36893-1 ISBN 978-3-319-17485-3 (eBook)
DOI 10.1007/978-3-319-17485-3

Springer Cham Heidelberg New York Dordrecht London

Softcover reprint of the hardcover 1st edition 2015

Printed on acid-free paper

Springer International Publishing AG Switzerland is part of Springer Science+Business Media
(www.springer.com)

Danksagung

Diese Arbeit entstand während meiner Zeit als wissenschaftlicher Mitarbeiter am Institut für Regelungstechnik der Technischen Universität Braunschweig. Für die schönen und lehrreichen Jahre trotz der vielen stressigen Phasen möchte ich mich bei allen Kolleginnen und Kollegen am Institut herzlich bedanken. Insbesondere gilt mein Dank Prof. Dr.-Ing. Markus Maurer nicht nur für die Betreuung dieser Arbeit und die mir dabei gegebenen Freiheiten, sondern insbesondere für die langjährige Verbundenheit und Förderung, die - angefangen mit einem Praktikum noch zur Studienzeit - deutlich über meine Zeit in Braunschweig hinaus geht. Letztlich hat genau dies meinen bisherigen Werdegang entscheidend geprägt. Prof. J. Chris Gerdes danke ich für die Übernahme des Koreferats und die schöne Zeit an der Stanford University, die ein Grundstein für diese Arbeit wurde. Der in den USA gesammelte technische wie auch nicht-technische Erfahrungsschatz hat viele Aspekte dieser Arbeit inspiriert oder erst ermöglicht. Prof. Dr.-Ing. Rolf Ernst danke ich für die Übernahme des Vorsitzes.

Insbesondere möchte ich mich auch bei meinen langjährigen Bürokollegen und Freunden Falko Saust, Horea Cernat, Asem Eltaher und Karsten Cornelsen sowie bei Bernd Lichte für die zahlreichen technischen und nicht-technischen aber immer sehr interessanten und oft erheiternden Gespräche bedanken.

Mein ganz herzlicher Dank gebührt auch allen Mitstreitern im Projekt MOBILE. Ohne eure Mithilfe wäre diese Arbeit nicht möglich gewesen. Vor allem gilt mein Dank den sehr zahlreichen Studenten, die ich in meiner Zeit als wissenschaftlicher Mitarbeiter betreuen durfte und die mit ihren einzelnen Beiträgen und oft weit überdurchschnittlichem Engagement und Einsatz den Grundstein für den Erfolg des Projekts gelegt haben. Bei meinen Kollegen im Projekt Torben Stolte, Bashar Ibrahim, Nico Selle und Sven Böhme möchte ich mich für die gute Zusammenarbeit bedanken und wünsche euch weiterhin viel Erfolg und Freude mit MOBILE - passt gut auf es auf! Unserer Werkstatt unter der Leitung von Andreas Rusniok gebührt mein Dank für die Unterstützung der zahlreichen "Spezialaufgaben", die unser Fahrzeug und die vorausgehenden Prüfstände maßgeblich mitgestaltet haben. Dennoch wäre all dies nicht ohne die Mithilfe und Kreativität unserer beiden Sekretariatsdamen Veronika Krapf und Stefanie Scheffer möglich gewesen. Herzlichen Dank an euch beide für die unzähligen Lösungen zu kleinen und größeren Problemen, die der Aufbau eines kompletten Fahrzeugs in Kombination mit den Verwaltungsrichtlinien einer Universität organisatorisch mit sich bringt.

DANKSAGUNG

Schließlich gilt mein ganz besonderer Dank meinen Eltern für die vielfältige Unterstützung und den persönlichen Rückhalt vom Studium angefangen bis hin zum Abschluss dieser Arbeit.

Nandlstadt, den 14.11.2014
Peter Bergmiller

Contents

Abstract

The design and testing of modern vehicle electronics are becoming more and more demanding due to increasing interdependencies among components and the safety criticality of tasks. The development towards Drive-by-Wire functionalities in vehicles with multiple actuators for vehicle control further increases the challenge. This thesis proposes approaches to address these challenges based on a vehicle level view and with a special emphasis on Drive-by-Wire systems. The interactions between components are explicitly taken into account and exploited to bridge the gap between the need to generate additional customer benefits and the effort to achieve functional safety.

In detail, a twofold approach is followed: on the one side, a toolchain to support efficient further development of novel functionalities for Drive-by-Wire vehicles is presented. The toolchain comprises appropriate software tools and scaled and full-scale experimental vehicles. On the other side, development towards functionally safe and flexible Drive-by-Wire vehicles is addressed by proposing a top-down designed architecture for vehicle electronics that is enabled by suitable mechanisms. The resulting goal achievement with regard to functional safety is evaluated based on a novel hierarchical approach.

The proposed architecture focuses on the reduction of classical hardware redundancies in the vehicle by exploiting the vehicle-wide functional redundancies among different actuators, which at the same time contributes to increasing customer benefits. The proposed mechanisms support this approach by improving vehicle monitoring and control both in long-term and short-term perspectives. This work refers to the latter as "tactical" mechanisms that detect and handle occurring failures, while "strategic" approaches aim to avoid critical situations a-priori and on a long-term basis. Strategically, wear and load of components in the vehicle are balanced across application domains, and the current state of the ego vehicle is monitored on a long-term basis by a self-representation system that assures knowledge of the vehicle about itself as a basis for decision making. For tactical failure handling, an efficient, traceable, and easily configurable probabilistic fault detection and handling algorithm is introduced. The algorithm implements the degradation concept of the vehicle by triggering appropriate actions according to the detected failures. Finally, the functional safety of the resulting overall system is evaluated using the mentioned hierarchical approach. The results demonstrate the benefits of the vehicle level design to achieve both functional safety and increase customer benefits with a reduced set of hardware components.

Kurzfassung

In modernen PKWs wird die Fahrzeugelektronik immer mehr zum Schlüsselelement in Hinblick auf neue Funktionalitäten jedoch auch zu einem wesentlichen Komplexitätstreiber mit entsprechenden Auswirkungen auf Produkt und Entwicklungsprozess. Die zu beobachtende Entwicklung hin zu Drive-by-Wire-Funktionen in Fahrzeugen, die mit neuartiger Aktorik (z.B. Hinterradlenkung) ausgestattet sind, verschärft die Problematik der Komplexitätsbeherrschung weiter. Um die damit einhergehenden Herausforderungen besser meistern zu können, wird in dieser Arbeit ein systemorientierter Ansatz vorgestellt. Dabei werden aus der Perspektive des Gesamtfahrzeugs heraus innovative Lösungsstrategien zur Unterstützung bei der Entwicklung funktional sicherer Kontrollsysteme für Drive-by-Wire Fahrzeuge vorgestellt. Besondere Beachtung findet dabei der Zielkonflikt zwischen erforderlichen neuen Funktionen für den Fahrzeugführer und gleichzeitiger Beherrschung der funktionalen Sicherheitsanforderungen unter Berücksichtigung von Kostenrandbedingungen.

Hierzu wird ein zweiteiliger Lösungsansatz vorgestellt: Einerseits wird eine Toolkette zur Unterstützung der Entwicklung neuartiger Fahrzeugelektroniksysteme implementiert. Diese beinhaltet sowohl Simulationselemente als auch Experimentalfahrzeuge einschließlich skalierter Modellfahrzeuge. Andererseits werden die Architektur eines Drive-by-Wire-Systems sowie notwendige Mechanismen für deren kosteneffiziente Realisierung anhand eines Experimentalfahrzeugs untersucht. Die entwickelte Architektur zielt insbesondere auf die Nutzung funktionaler Redundanzen auf Gesamtfahrzeugebene ab. Hierdurch können kostenträchtige klassische Redundanzmechanismen auf Komponentenebene bei gleichzeitiger Einhaltung der Sicherheitsziele teilweise ersetzt werden. Die als Grundlage für den Architekturansatz entworfenen Mechanismen werden in dieser Arbeit in langfristig und vorausschauend operierende "strategische" Mechanismen und kurzfristig intervenierende "taktische" Systeme unterteilt.

Zur strategischen Vorbeugung kritischer Situationen werden Verschleiß und Lastzustände zwischen unterschiedlichen Aktoren im Fahrzeug ausgeglichen. Zusätzlich wird der aktuelle Zustand des Eigenfahrzeugs im Rahmen eines Systems zur Selbstrepräsentation langfristig überwacht. Auf Basis dieses Wissens kann das Fahrzeug automatisch hoch sicherheitskritische Entscheidungen in Bezug auf den Einsatz einzelner Aktoren treffen. Taktische Maßnahmen überwachen den Zustand des Eigenfahrzeugs kontinuierlich auf Basis lokaler Messungen und intervenieren im Fehlerfall innerhalb weniger Millisekunden. Hier wird als wesentliche Komponente ein probabilistischer und echtzeitfähiger Diagnoseansatz vorgestellt, der entsprechend des aktuell wahrscheinlichsten Fehlers unterschiedliche Aktionen im Rahmen des Degradationskonzepts ausführen kann. Um die Erreichung funktionaler Sicherheitsziele mittels der vorgestellten Ansätze zu evaluieren, wird eine hierarchisch strukturierte Herangehensweise zur Gesamtfahrzeuganalyse vorgestellt, die funktionale Redundanzen und Wechselwirkungen im Gesamtfahrzeug berücksichtigt. Anhand der Ergebnisse werden die Vorteile eines gesamtfahrzeugorientierten Ansatzes bei der Entwicklung neuer Fahrzeugelektroniksysteme deutlich.

Part I: Introduction

This part of the thesis describes the need for new approaches, architectures, methods, and tools in the field of vehicle electronics with a focus on Drive-by-Wire applications. Therefore, an assessment of the current standards and development processes in the industry is given. Finally, the structure and the contribution of this thesis are introduced.

1

Vehicle Electronics: A Challenge for the Automotive Industry

"To improve is to change; to be perfect is to change often."

Winston Churchill

Increasing complexity of modern vehicles

In the Federal Republic of Germany, more than 719,535 people were employed in the automotive industry in 2011. The industry branch generated an overall turnover of about 351 billion Euros according to the German Federal Transport Authority (German: Kraftfahrt-Bundesamt), and 20 billion were spent on research and development [BMWi, 2012]. The production of six million cars a year made Germany the number one car selling country in Europe and the fourth biggest car selling country in the world [OICA, 2012]. This shows that the automotive industry contributes hugely to the national output of Germany and thus is vital for the further economic success of the country [Legler et al., 2009]. At the same time, the complexity of modern vehicles is continuously increasing and vehicle development is becoming increasingly challenging since additional and more complex functionalities from different domains (ergonomics, entertainment, emission control, etc.) are being demanded by the customer [Sangiovanni-Vincentelli, 2007] and the legislator. Adaptations and extensions of the electrics and electronics (EE) of the vehicle – especially the integration and interaction of previously independent functions – significantly contribute to meeting these demands [Arbitmann et al., 2011; Sinha, 2011; Pruckner et al., 2012]. Thus, the number of interconnections and interdependencies within the EE system is rapidly increasing on the functional and hardware level [Schäuffele and Zurawka, 2013], which greatly heightens vehicle complexity.

New functionality vs. safety

In parallel, the functional safety of the vehicle has to be ensured. This generates highly conflicting goals of additional functionality and proof of functional safety[1]. With electric vehicles joining the market, meeting both types of goal becomes even more challenging. Depending on the drive train structure, electric vehicles offer powerful ways to influence vehicle handling, e.g., with torque vectoring[2]. These

[1]According to ISO 26262, functional safety refers to the "absence of unreasonable risk due to hazards caused by malfunctioning behavior of E/E systems" [ISO 26262-1, 2011, p. 7].

[2]Torque vectoring refers to an approach where individual wheels of a vehicle are driven with

P.J. Bergmiller, *Towards Functional Safety in Drive-by-Wire Vehicles*,
DOI 10.1007/978-3-319-17485-3_1

systems are controlled "by-Wire"[3] and completely automatically through the EE system. Thus, a failure[4] of EE components or a faulty decision by a relevant controller can render the vehicle uncontrollable for the driver, which may lead to fatal crashes as recently revealed in a law suit against Toyota [Benjamin, 2014]. Therefore, by-Wire control has long been the subject of research, but customer needs are now triggering its gradual introduction into series vehicles for highly safety critical applications, e.g., for rear-wheel steering [Pruckner et al., 2012], which extends the already accepted use for "Throttle-by-Wire". Consequently, further profound development in the field of vehicle electronics is crucial not only for the German automotive industry to maintain its leading global position based on innovative functionalities, but also to make driving safer. This necessity is strongly supported by a technical survey issued and funded by the Federal Ministry of Economics and Technology in Germany (BMWi) [Bernard et al., 2010]. The survey especially demands fundamental reconsideration of the established electronics architecture in series vehicles in terms of performance, reduction of complexity, and functional safety. These are regarded as key factors affecting economic success and further technical progress in the field of vehicle electronics.

The vehicle as one system

Turning from the classical perception of the vehicle as a number of interconnected parts to a more holistic functional perception of the overall vehicle can be a way to address these key factors. This automotive systems engineering approach is followed at the Institute of Control Engineering at the TU Braunschweig [Maurer, 2013] and is also starting to be adopted by some companies [Abele, 2012; Papadopoulos et al., 2001]. The approach especially focuses on interactions across different systems in the vehicle, identification of synergies and conflicts, and functional goals on the vehicle level, including safety goals. With the decreasing vertical range of manufacture of OEMs[5], these investigations have become a core task of the OEM [Legner et al., 2009, p. 7]. Methods, tools, and designs resulting from the vehicle level investigation constitute useful extensions to the existing approaches and thus are investigated in the MOBILE project at the TU Braunschweig with a special focus on Drive-by-Wire applications. The following chapter will introduce the contributions of this work and relate them with deficits in the typical processes and methods in used in the automotive industry.

individual drive torques. When driving the wheels on one side of the vehicle with a different torque than the wheels on the opposite side, an additional yaw moment is generated. For further information including evaluation of safety criticality see, e.g., Euchler et al. [2010].

[3]In the automotive field, "by-Wire" control means that actuators in the vehicle are controlled purely electronically without any mechanical or hydraulic linkage between the actuator and the driver. Well known examples include Brake-by-Wire, Steer-by-Wire, or Shift-by-Wire.

[4]ISO 26262 defines failure as the "termination of the ability of an element to perform a function as required" [ISO 26262-1, 2011, p. 7].

[5]Original Equipment Manufacturer, e.g., AUDI AG or BMW AG for the automotive domain

2

Redesigning Vehicle Electronics: A Systems Engineering Approach

> "The world we have made as a result of the level of thinking we have done thus far creates problems we cannot solve at the same level of thinking at which we created them."
>
> *Albert Einstein*

This thesis follows an Automotive Systems Engineering approach as introduced by Maurer [2013] to address existing and arising problems with individual components in the vehicle by taking into account vehicle-wide cross dependencies and interactions. This way, new solutions to existing problems can be found by a targeted analysis and reorganization of existing systems, and opportunities to realize novel functions become obvious. The thesis especially focuses on novel approaches for Drive-by-Wire systems. To further clarify the goals and contributions of this thesis, a short assessment of typical development processes in the automotive field (Sec. 2.1) is given. Complementary, important standards and legislative demands relevant to this work are referenced, required terminology is defined, and the contribution of this work is put into relation to existing approaches. An overview of the structure of this document (Sec. 2.2) and a short introduction of the project background (Sec. 2.3) conclude the thematic identification.

Automotive systems engineering

2.1. Current Practices: Process Models and Methods

Processes relied on in industry play a vital role in mastering the complexity of the development of modern vehicles [Bender, 2005, p. 1]. The product development process is defined as a "system of defined steps and tasks such as strategy, organization, concept generation, marketing plan creation, evaluation, and commercialization of a new product. It is a cycle by which an innovative firm routinely con-

Process models

P.J. Bergmiller, *Towards Functional Safety in Drive-by-Wire Vehicles*,
DOI 10.1007/978-3-319-17485-3_2

verts ideas into commercially viable goods or services" [BusinessDictionary, 2012][1]. Appropriate structuring and control of these processes significantly impacts the economic success of a company [Mehrle et al., 2012; Ward and Sobek, 2014]. For that reason, a multitude of process and phase models[2] have been developed and applied [Benington, 1983; Hammerschall, 2008]. These models support planning, control, monitoring, and execution of product development by introducing reference procedures. Nowadays, multiple proprietary[3] and standardized process models are applied, such as the Unified Process [Jacobson et al., 1999], the OPEN Process [Graham et al., 1999], or the V-Model in its different derivatives (V-Model 92/97/XT) [Kuhrmann et al., 2011]. Further models and detailed comparisons can be found in Hammerschall [2008]. Due to its widespread usage in the automotive domain, the research in the MOBILE project refers to the V-Model as a basis for the thematic identification.

The V-Model

The V-Model XT 1.4 (XT standing for Extreme Tailoring) structures product development both for the orderer and the contractor [V-Modell, 2012]. Therefore, it includes various steps from project approval and system development to the project end. For smaller projects, the V-model XT can be adapted to reduce project overhead. Prominent institutions and international standards refer to the V-model and thus further push its development. For example, the Federal Republic of Germany relies on the V-model for IT projects [BMI, 2010][4], and the important automotive standard ISO 26262 "Road Vehicles – Functional Safety" relies on the V-model to structure the development process.

The V

In this thesis, only the part of the V-model dealing with technical product development is referenced. This "core" of the V-model, first developed by Rook [1986], was successively supplemented in follow-up versions to create the current V-Model XT and is structured as given in Fig. 2.1. Based on the description of the requirements, the overall system is specified and the (functional) architecture of the system is derived. Then, the system is detailed in sub-systems and modules, which are specified and finally implemented (coding). Afterwards, the hierarchical path is followed back up to the final acceptance test of the system. At each hierarchical layer, the implemented systems are tested for compliance with the previously defined requirements.

V-model for mechatronic systems

All process models referenced before, including the V-model, originate from the field of software engineering and therefore are targeted primarily at software development. When mechatronic systems for vehicles are developed, new challenges arise

[1]A similar definition can be derived, if the perception of a process as "a series of things that are done in order to achieve a particular result" by [Wehmeyer, 2005, p. 1202] is combined with the definition of "new product development" by [Collins, 2013] as "the process of developing new products for the market". For this work, the formulation by [BusinessDictionary, 2012] is found more appealing and therefore referenced.

[2]Phase models describe only the top-level development flow, e.g., analysis, design, implementation, whereas process models include additional information, e.g., which results are expected in detail and who is responsible for a task [Hammerschall, 2008, p. 31].

[3]specialized for a specific organization; not further considered here

[4]Federal Ministry of the Interior of Germany

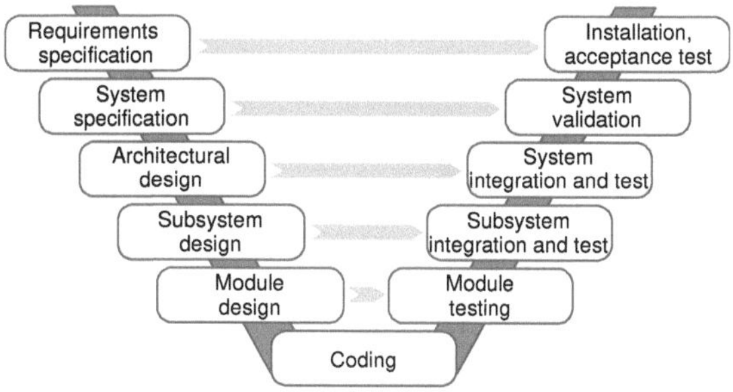

Figure 2.1.: V-model based development; figure according to Redmill [1997]

due to the close coupling of newly designed mechanics, EE components, and software. These aspects are not considered sufficiently in the V-model as introduced. Within the EQUAL research project, an accordingly adapted "3-Ebenen-Modell" (3-Layer-Model) was developed [Bender, 2005, p. 45]. Figure 2.2[5] outlines the resulting V-shaped process with the three disciplines (mechanics, EE, and software) executed in parallel. Each discipline performs a V-shaped development process, but the 3-Layer-Model synchronizes and merges tasks at relevant steps during product development. This indicates the increase in complexity and interdependencies between the different fields during product development. Especially during the testing phases at system and subsystem level, products from multiple disciplines must be available to allow component and integration tests. One missing part is enough to delay the overall process. For the same reason, testing at system or vehicle level under realistic conditions during the early phases of product development is barely possible. Still, such early iterations are necessary [Hermes and Schultze, 2009; Maurer, 2012] especially in pre-development and research and to support the Automotive Systems Engineering approach. Accordingly, methods[6] and tools to evaluate changes applied at the system level during early development phases have to be provided. The V-Model itself does not specify any methods for the individual development steps [Wieczorrek and Mertens, 2011]. On the one side, this contributes to the success of the V-Model in different domains. On the other side, methods suitable to support certain process steps have to be found independently. As this work introduces and applies amongst others such a method, a brief assessment of typical elements of a method and the relation to processes, principles, and heuristics is given. The differentiation is performed following Hammerschall [2008].

The core of a method is a process, which is defined as a "series of things that are done in order to achieve a particular result" [Wehmeyer, 2005, p. 1202]. Thus, a detailed process provides a suitable guideline for what has to be done to achieve a Methods

[5]The numbers and highlighting in the figure are for later reference.

[6]A method is defined as "a way of proceeding or doing something, esp. a systematic or regular one" [Collins, 2013]. During a development process, (multiple) methods can be applied to achieve necessary results [Hammerschall, 2008]. Further details are given in the next paragraph.

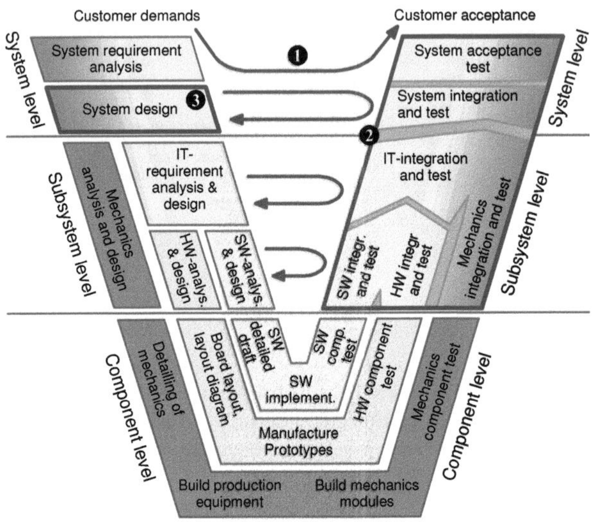

Figure 2.2.: "3-Ebenen-Modell" (3-Layer-Model) [Bender, 2005, p. 45][8]

certain result. But, to maximize the number of possible fields of application for a method, the process specified by a method can deliberately be incomplete. Missing parts allow adaptation of the method to multiple fields of application. In order to ensure that these adaptations focus on the purpose of a method, methods can additionally specify principles[7]. The principles describe general conditions for the process. Well known examples that are also applied in this work are the "top-down" or the "bottom-up" principle [Hammerschall, 2008, p. 61]. Additionally, supportive heuristics can be included in a method. Heuristics can assist in "solving problems by finding practical ways of dealing with them learning from past experience" [Wehmeyer, 2005, p. 730]. Figure 2.3 summarizes the relations between method, principle, and process in a modified UML structure diagram according to Gnatz [2005]. The additional arrows and text markings indicate the meaning of the links and the reading direction.

Standards, legislation

As complex mechatronic systems that have been developed according to the outlined automotive development processes and using suitable methods are increasingly taking over safety critical tasks (compare Cha. 1), they are attracting increased public interest and also, therefore, normative and legislative regulations. These additional requirements complement the challenges for system level development outlined so far and will be referenced in the appropriate sections. Sec. A.1

[7] A principle is defined as "an underlying or guiding theory or belief" [Collins, 2013].

[8] The permission to print Fig. 2.2 in the proposed English translation from the German original version was received from Klaus Bender at January 31st 2013.

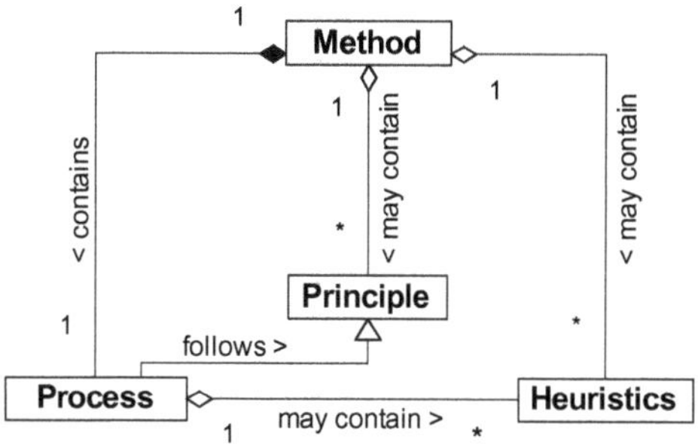

Figure 2.3.: Relations between method, process, principle, and heuristics

provides a summary of important standards and regulations that were taken into account and briefly introduces the relevant standardization institutions.

2.2. Structure and Contribution of This Thesis

The given glance at current development processes in the automotive industry clarified the challenges that exist during the development of mechatronic systems. Drive-by-Wire systems are especially effected due to the high safety criticality of the vehicle control task. Accordingly, this thesis proposes methods and designs that assist with achieving functional safety in Drive-by-Wire vehicles by exploiting the vehicle level context. Challenges that are hard to address efficiently at component level are handled at vehicle level. Still, the review of the V-model development process revealed that especially iterative work on higher hierarchical layers, such as vehicle/system layer, is barely supported by methods and tools nowadays, which necessitates to consider these aspects in this thesis. As will be outlined, the resulting contributions are mostly not restricted to Drive-by-Wire systems but can be transferred to other fields of application of vehicle electronics.

Tools and solutions

Specifically, this work introduces (a) a toolchain for development and evaluation of interconnected vehicle electronics at system level, and (b) an architecture for a highly flexible Drive-by-Wire vehicle. The architecture explicitly targets system wide interaction of components to meet safety goals. It is flanked by an appropriate safety concept relying on novel monitoring and intervention mechanisms and an appropriate method to assess functional safety at vehicle level. The development of the toolchain and the architecture are closely linked: on the one side, the architecture is implemented on a real experimental vehicle using the toolchain. On the other side, the experimental vehicle, which features the flexibility enabled by the novel architecture, becomes a core part of the toolchain, as it provides the platform for test runs in a real vehicle. In summary, this work will paint a "big picture" of a novel approach based on the proposed architecture and the developed vehicles, but will also enrich this picture with some details in important parts. Basically, the presented vehicles provide huge possibilities for investigations and research, and the mentioned details focus on enabling novel approaches or demonstrate the performance of the architecture and the vehicles using an example application.

Toolchain

The mentioned toolchain enables the proposed vehicle level design approach. Essentially, it supports easy evaluation of algorithms and electronics operating at vehicle level in a real vehicle at early development stages. Therefore, several key aspects are covered:

- Virtual testing is facilitated by a simulation environment including vehicle dynamics models and a virtual driver that can easily be adapted to drive various vehicles with different actuator set-ups.
- The algorithms developed in simulation can quickly be ported onto microcontrollers or onboard computers on experimental vehicles. Suitable operating systems for the microcontrollers and a link to the Mathworks Simulink environment are provided. Both a scaled and a full-scale experimental vehicle are available, which support safe stepwise testing of safety critical algorithms.
- To visualize data in the vehicles, a server-client based system facilitates distributed access to all vehicle data from within and outside the experimental vehicle. The system supports online diagnostics and allows the user to individually configure the look of the dashboard.

This toolchain assists the developer during quick iterations on higher hierarchical levels of the V-Model that are so far little covered by existing tools. Figure 2.2 highlights the relevant steps in the V-Model for mechatronic components. Front-loading of component and integration tests under real driving conditions (❷, Fig. 2.2) is supported. For example, an electric drive unit and its integration with other completed parts can be investigated in the experimental vehicle before the remaining vehicle exists. Novel architectural approaches can be evaluated due to the modularity and reconfigurability of the experimental vehicle. Another special field of application targets early evaluation of how well customer needs can be met by a new functionality with relevance to vehicle handling (❶, Fig. 2.2). Therefore, the powerful experimental vehicles can simulate the dynamic behavior of vehicles that do not yet exist. Thus, the longitudinal and lateral handling of virtual vehicles can be evaluated on the test track in a way that is similar to the approach by Cornelsen et al. [2011] for longitudinal dynamics. More details on the toolchain and the special contribution of the experimental vehicles will be given in Cha. 3 and Cha. 4.

A novel system architecture

As mentioned, as a part of the toolchain, the flexible experimental vehicles are themselves subject to research. Based on requirements posed by the toolchain and further constraints due to costs and availability of components, a novel architecture is introduced for the EE system of these vehicles. This architecture supports flexibility while providing the required degree of functional safety (Cha. 5). A

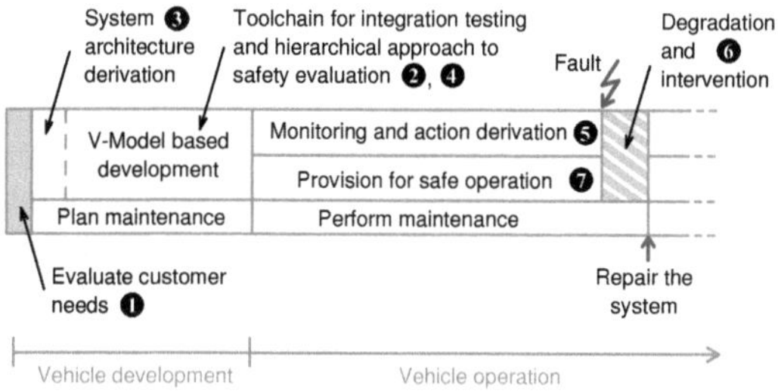

Figure 2.4.: Simplified vehicle life cycle with associated contributions of this thesis

primary objective of the architecture is to exploit functional redundancies across different types of actuators for safety purposes. Therefore, the structure of the system is introduced top-down starting from the vehicle level and relying on different views (functional, hardware, software, etc.) of the overall system. This structured approach supports system design by explicitly questioning traditional system structures in vehicles (❸, Fig. 2.2).

Mechanisms and methods

As enabling mechanisms for the proposed architecture, both preventive long-term approaches and methods for immanent intervention in case of failure are introduced. This work refers to the first group as "strategic" (Cha. 8) and to the second group as "tactical" (Cha. 7) mechanisms. Strategic measures aim to avoid safety critical situations by anticipative planning and intervention exploiting the flexibility of the vehicle. Tactical measures perform online monitoring and derive actions in case of failure so that at least degraded vehicle operation is guaranteed until a safe state can be achieved. As core mechanisms in the field of strategic measures, this work introduces an *online-optimization system* that balances load and wear among different types of actuators, and an *ability based self-representation* for the vehicle that continuously monitors the health state of the vehicle and identifies maneuvers and actions that can no longer be performed by the vehicle. This knowledge is vital for the automated decision making of the vehicle on usage of its actuators, but also as a basis for later development of autonomously driving vehicles with environmental perception. Tactical aspects covered in this thesis include an *efficient probabilistic approach* to online monitoring that supports high flexibility in the software system and a short assessment of *stability control* using four-wheel steering and torque vectoring to compensate for failures of individual actuators. For the overall system, a safety evaluation is provided (Cha. 9). Therefore, a novel method for *tailored hierarchical safety analysis* (Cha. 6) was developed that especially focuses on exploitation of functional redundancies and targeted safety evaluation.

Summary of Contributions

As stated in the objectives, the toolchain and the introduced architecture including the developed mechanisms help to close critical gaps in the development process of powerful and highly-safety critical vehicle electronics. Figure 2.4 associates the main contributions of this thesis to a simplified vehicle life cycle, with a focus on the introduced aspects. The proposed toolchain can assist in evaluating customer benefits from novel functions (❶, Fig. 2.4). If a function is found suitable, the introduced architecture for flexible vehicles (❸, Fig. 2.4) or individual aspects of it may serve as an input for the development of the EE system of a vehicle. Thus, the reconsideration of existing domain-oriented system architectures can be triggered. During system development, the toolchain assists by enabling early testing during iteration loops at the vehicle/system/subsystem level, which are currently not supported sufficiently by available tools (❷, Fig. 2.4). The hierarchical approach (❹, Fig. 2.4) to safety analysis provides an estimate of the functional safety of the vehicle based on the first development steps, and can provide increasingly detailed results if the inputs of the safety analysis are continuously complemented in the following steps. During vehicle life, tactical safety measures, such as the monitoring and action derivation algorithm (❺, Fig. 2.4), ensure cost efficient detection and treatment of failures that occur while driving. If a failure is detected, control algorithms for vehicle dynamics contribute to the degradation concept (❻, Fig. 2.4). First research results on cross-actuator failure compensation are considered and provide an outlook on what could be possible in future. Strategic mechanisms (❼, Fig. 2.4) continuously operate while driving and reduce the probability of the occurrence of risky situations that might result in safety critical failures. Therefore, online optimization unburdens overloaded actuators, and a skill-based self-representation, e.g., prevents execution of maneuvers that are no longer regarded as safe, judging from the current vehicle state. The benefits of the individual outlined aspects will be revisited with the example of the constructed experimental vehicle "MOBILE". As a graphical guideline, Fig. 2.5 summarizes the structure of the thesis.

2.3. Project Background and Constraints

For a better understanding of decisions made in this work, a brief outline of the general conditions under which the MOBILE project presented in this thesis was carried out concludes the introduction. One of the main influencing factors is the "university only" character of the project. The project was not primarily funded by any partner from industry, which led to great freedom in decisions and design but also very definite financial limits. The main goals of the project, apart from scientific aspects, were (a) to build up know-how in various fields related to vehicle electronics, (b) to introduce a vehicle level perspective into electronics design driven by challenges observed in the projects with industry, and (c) to provide a basis for future research in these fields. The decision to construct the fully custom-built vehicle MOBILE for that purpose was inspired by the positive experiences with the vehicles "X1" and "P1" built by Gerdes at the Dynamic Design Lab (Stanford University, USA). Building the vehicle forces intense consideration of all relevant aspects from the component level up to the vehicle level ("We will learn something

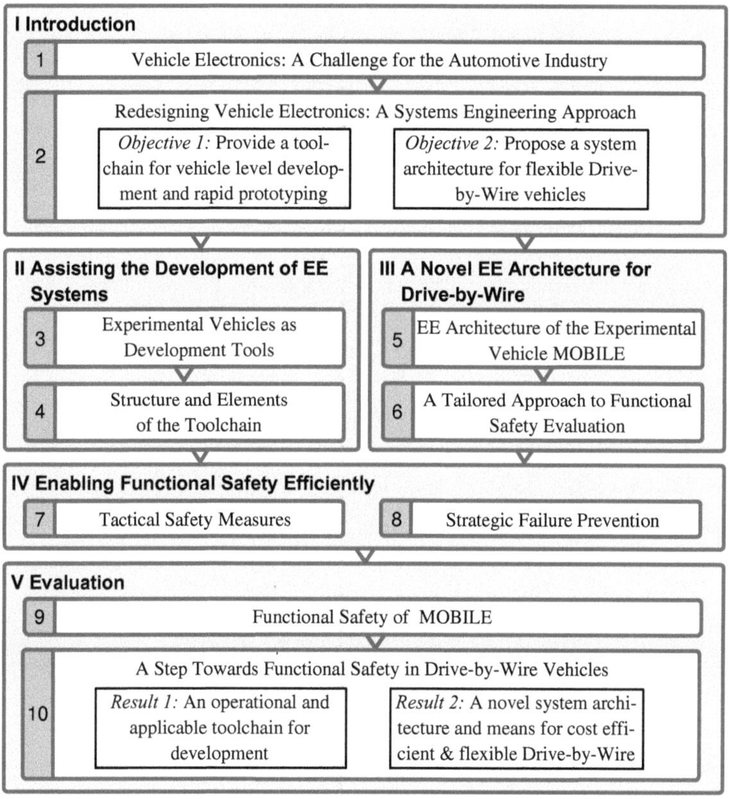

Figure 2.5.: Graphical table of contents

from it", Gerdes).

The educational aspect of the project results in multiple self-designed systems. This facilitates full accessibility for students and researchers, and supports the accumulation of critical know-how in novel fields, such as Drive-by-Wire and electromobility, at the university. Accordingly, the project was hugely supported by numerous students writing their bachelor's or master's theses (see bibliography), which were combined into an overall operational system. In summary, the university driven character of the project facilitates "out-of-the-box" thinking and novel approaches but also guides decisions on whether to purchase or design components.

Part II: Assisting the Development of EE Systems

Development and testing of novel, vehicle-wide operating electronics is becoming increasingly challenging. This part describes the need for appropriate tools and introduces a suitable toolchain developed in the MOBILE project.

3 Experimental Vehicles as Development Tools

"It doesn't matter how beautiful your theory is, it doesn't matter how smart you are. If it doesn't agree with experiment, it's wrong."

Richard P. Feynman

Sec. 2.1 introduced the V-model as a reference for the automotive development process and pointed out the special challenge of early iterative testing of complex mechatronic systems. These systems feature multiple dependencies from the other components of the vehicle, driver inputs, or data acquired from the vehicle environment. Thus, proper tools are needed that provide realistic inputs to the system under test [Düser, 2010]. In particular, algorithms intervening in vehicle control require early and sufficient testing during real test drives [Bock, 2008]. Accordingly, development and test systems that are based on real vehicles will more and more be needed to extend existing tools especially for iterations in early phases of product development [Hermes and Schultze, 2009; Maurer, 2012].

Commonly used tools

So far, closed-loop tests, also referred to as "X-in-the-Loop" tests [Reitze, 2004], are commonly performed to evaluate the proper operation of a system under test in dynamically changing environmental conditions. Traditionally, these approaches stimulate the system under test (represented by the "X") with inputs generated by a virtual environment. The reactions of the system under test are fed back to the virtual environment and trigger reactions forming the "loop". Well-known examples of this type of test are Model-in-the-Loop (e.g., [Plummer, 2006]), Software-in-the-Loop (e.g., [Chen et al., 2008]), or Hardware-in-the-Loop (e.g., [Lu et al., 2007]) tests. Due to the virtual environment, these approaches have significant advantages for testing, e.g., in terms of reproducibility and effort and thus are frequently relied on. Further simulation based approaches including driving simulators are summarized by Düser [2010] and Slob [2008]. Still, the mentioned approaches often lack precision for detailed investigation of complex systems in vehicles [Düser, 2010]. The deficiencies result from the simulated components of the environment: during early phases of product development, simulation can often provide sufficiently good

P.J. Bergmiller, *Towards Functional Safety in Drive-by-Wire Vehicles*,
DOI 10.1007/978-3-319-17485-3_3

imitation of real processes to appropriately stimulate the system under test. But as development progresses and the need to detail the environment including the driver behavior increases [Asano et al., 1991], the effort required to implement the simulation increases exponentially, while the quality of results can still barely meet the required level ([Heißing, 2002] according to [Düser, 2010, p. 22]). Thus, real test runs are mandatory.

Need for flexible vehicles

As this work focuses on the investigation of highly safety-critical electronics, which intervene in vehicle control and depend both on driver inputs and environmental influences, appropriate experimental vehicles for real test runs are required. As the ongoing research at the Institute of Control Engineering will need such vehicles in multiple research projects for different research purposes, a flexible vehicle should be constructed instead of multiple vehicles for specific purposes [1]. Emery [1998] working for the National Highway Traffic Safety Administration in the USA and confronted with a similar challenge describes this demand as follows:

> *"It was recognized that testing costs could be greatly reduced and testing efficiency could be greatly increased if one could more easily change the test vehicle performance parameters. [...] The study[a] concluded that the design and construction of such a vehicle was technically feasible and that such a testbed vehicle would be extremely valuable for conducting conventional as well as high technology safety research." [Emery, 1998, p.1]*
>
> [a]Author's note: carried out by the National Highway Traffic Safety Administration on use cases and structures of flexible vehicles

Goal of this chapter

Consequently, the following section provides an overview of highly flexible experimental vehicles that have so far been constructed by other research groups and companies as test and development tools. The section focuses on vehicles that are intended as tools to evaluate different mechanical designs relevant to vehicle dynamics, different actuator set-ups, or different control algorithms. Prototype vehicles dedicated to a sole research project or intended to demonstrate a specific technology or design to the public without further usage of the vehicle in other research projects are not considered. The "lessons learned" from the introduced vehicles serve as input for the development of MOBILE. In particular, the best ways to achieve flexibility of the individual vehicles will be investigated.

3.1. Overview of Research Vehicles

RWTH Aachen

Vehicle dynamics, controllability[2], and driving comfort were already research topics before electronics started to be featured prominently in vehicles. One of

[1]In aerospace, similar flexible systems have long been used, e.g., to simulate the dynamics of not yet existing airplanes [Kauffman et al., 1949; DLR, 2014].

[2]Controllability refers to the "ability to avoid a specified harm or damage through the timely intervention of the persons involved, possibly with support from external measures" [ISO 26262-1, 2011, p. 4].

Figure 3.1.: "Experimental-Handling-Fahrzeug", RWTH Aachen [Fuhrmann, 1980][4]

the first mechanically highly flexible vehicles built for such investigation was the "Experimental-Handling-Fahrzeug[3]" (EHF) built at the RWTH Aachen in 1978 [Fuhrmann, 1980]. This modular vehicle was built based on a square tube frame featuring three modules: a front module, a center module containing the driver's seat and a powerful engine (a Mercedes E 280 engine) for front or rear-wheel drive, and a rear module (Fig. 3.1). The modular design facilitated evaluation of different suspension systems and modification of important characteristics, such as camber, caster, king-pin angle, or track-width. Also, damping rates could be modified mechanically. The vehicle was based on the ideas of Dr. Bernd Heißing, who later supervised the construction of a similar vehicle at the TU München, as will be pointed out later. The construction plans of the Experimental-Handling-Fahrzeug were sold to Goodyear and Mercedes Benz.

Mercedes Benz

Zomotor [1991] introduces a research vehicle by Mercedes Benz for investigation of vehicle dynamics. The vehicle featured a similar structure to the modular vehicle built at the RWTH Aachen. Additional weights could be placed on the front and rear modules, and the height of the center module relative to the front and rear module could be modified. Like the research vehicle of the RWTH Aachen, the experimental vehicle did not feature bodywork. Consequently, all relevant components were readily accessible, which is especially valuable for investigation of vehicle dynamics [Zomotor, 1991, p. 299].

Porsche AG

Another prominent early prototype that focused on the mechanical design of a vehicle is described by Nötzli [1987]. Porsche built the "Porsche Experimental Prototype" that was used to investigate vehicle dynamics, engine management, and aerodynamics. The base square tube frame again consisted of three main modules: a front and a rear axle module that were mounted on a central module that contains the passenger seats. An engine module could be placed in any of the three modules and be connected to either the front, the rear, or both axles. The modular hull for investigation of aerodynamics particularly distinguished the Porsche Experimental Prototype from other vehicles.

[3]German "Fahrzeug" means vehicle

[4]The permission to print the photograph was received from the Institut für Kraftfahrzeuge RWTH Aachen at February 1st 2013.

Figure 3.2.: Research vehicle "X1", Stanford University (Dynamic Design Lab)

Nissan

Nissan constructed a modular vehicle that added flexibility to actuator control [Asano et al., 1991]. Nissan's goal was to "integrate human factors into control based on human response to vehicle dynamics" [Asano et al., 1991, p. 1], which so far had not been possible with the existing development tools. The "In-Vehicle Simulator" (IVS) consisted of five modules: the front, rear, and cabin modules that were featured in the other experimental vehicles were complemented by two extension modules that allow the wheelbase of the vehicle to be modified (2.5m, 3m, or 3.3m). Similar to the vehicle by Mercedes, mountable weights allowed to adjust the weight distribution. An active suspension system in combination with all-wheel steering and the possibility for front-, rear- and all-wheel drive facilitated investigation of various handling characteristics.

TU München

At the TU München, the "Experimental Handling Vehicle" (EHV) was built in a student driven project with support from industry [Meyer-Tuve et al., 2007] and is still in use. Unlike the projects mentioned so far, the vehicle is primarily intended as a carrier vehicle for novel mechanical or mechatronic modules and measurement equipment. This hugely aids research, as the mechanical package in series vehicles is dense and allows barely any modifications. Additionally, interfaces and functions of vehicle electronics are usually inaccessible to researchers. As outlined by Meyer-Tuve et al. [2007], the EHV is powered by a 254kW combustion engine mounted in the middle of the vehicle. Additionally, the axle kinematics can be adjusted offline.

Stanford University

The experimental vehicles "P1" and "X1" built at the Dynamic Design Lab of Gerdes at the Stanford University feature significant capabilities both in actuators and mechanical modularity [Beal and Gerdes, 2010; Bergmiller et al., 2008]. Both vehicles were custom built based on a square tube frame. The research vehicle P1 features a Steer-by-Wire system for the front axle with a force-feedback module. The steering system allows independent control of the left and right tire. The vehicle is driven by two electric motors connected to the rear left and rear right wheel each. Thus, torque vectoring is supported. The successor vehicle X1 (Fig. 3.2) adds mechanical modularity and can be split up in a front, middle and rear module. In its current version, X1 features electric rear-wheel drive, all-wheel steering with a force-feedback module, and individually controllable braking units for each wheel.

Complementing the above list of full-scale vehicles, Trächtler's (Universität Paderborn) go-kart-like "Chamäleon" vehicle has to be mentioned [Leppin and Wittmann, 2010]. The Chamäleon is a lightweight vehicle (280kg) with a top speed of 60km/h. The vehicle is equipped with four identical wheel modules. Each wheel module features three electric motors. A 2.2kW drive motor facilitates independent drive of each wheel. Two additional motors allow individual steering of each wheel and actuating the suspension system. The vehicle is used for real test runs and can be connected to a simulator for virtual test drives [Kreft et al., 2010]. Then, the driver sits in the vehicle, but gas commands are forwarded to the simulation environment instead of the vehicle actuators.

Universität Paderborn

A vehicle similar to "Chamäleon" was constructed at The Chinese University of Hong Kong and is used for investigating novel parking strategies [Qian et al., 2011]. The "4-wheel independent steering robot" (4WIS) features four wheel steering by independently controllable actuators and four in-wheel motors but no brakes. The small vehicle is intended for low speed applications.

The Chinese University of Hong Kong

The DLR's go-kart-like "ROMO" vehicle features four in-wheel motors with torques up to 160Nm per wheel and a steering angle of approx. 90 degree in one direction to facilitate driving orthogonally to the vehicle orientation [Brembeck et al., 2011]. Additionally, the vehicle features a hydraulic braking system taken from a go-kart which can be actuated by an electric motor. In summary, the vehicle has limited dynamics but can perform various complex maneuvers based on the in-wheel motor set-up and the huge steering angles. This set-up is extended by an environmental perception system to demonstrate the opportunities available when driving automatically with a highly flexible vehicle.

DLR

At the Kanagawa Institute of Technology, Abe's research group built an experimental vehicle in the minicar size with a wheel base of approx. 2.05m and a weight of 612kg. The vehicle features all wheel drive based on in-wheel motors similar to the DLR ROMO vehicle. Additionally, each wheel can be steered independently [Abe et al., 2013].

Kanagawa Institute of Technology

Extending the above mentioned custom-built vehicles, relevant projects that modified series vehicles with additional actuators to generate flexible platforms for evaluation of vehicle dynamics will be pointed out. The first approaches were taken by General Motors. The company built a research vehicle for investigation of handling characteristics in the 1970s. The vehicle featured electrohydraulic front and rear-wheel steering and a steering feel system to evaluate directional control characteristics of different vehicle set-ups [McKenna, 1974].

General Motors

Similarly, the Melbourne University introduced the "Variable Characteristic Car" (VCC) with front-wheel steering based on electrohydraulic actuators and a mechanical clutch system as a fall-back layer for steering [Dorey et al., 1980]. These early projects especially had to cope with challenges related to actuator design and the necessary electronics to perform real-time control.

Melbourne University

NHTSA

Lee et al. [1997] present an analysis of the capabilities of a "Variable Dynamic Testbed Vehicle" with regard to emulation of different vehicle handling characteristics. The study was conducted by the National Highway Traffic Safety Administration (NHTSA) to derive the set-up of a flexible vehicle that obviates the need to build specific prototypes in each research project. The simulation-based study concluded that a vehicle with all wheel steering, steering feel emulation, an active suspension system, Brake-by-Wire and Throttle-by-Wire can sufficiently emulate the dynamics of a wide range of other vehicles. As a result, a vehicle addressing the above needs was built based on a Ford Taurus and components from General Motors and various other automotive suppliers. The individual actuators were controlled by a single central control unit. Hydraulic and mechanical back-up systems for brakes and steering ensured safe operation of the vehicle.

Volvo, Chalmers University

Johannessen [2001] introduces the "SIRIUS 2001", a Lotus Super 7 replica that was modified by Volvo in cooperation with the Chalmers University of Technology in Sweden. The modified vehicle features all-wheel steering with independently controllable wheels. Driver commands are acquired by a steering input unit that can be mounted in front of the left or the right seat. Additionally, electric pumps control the pressure of each brake caliper and thus facilitate Brake-by-Wire. A central combustion engine drives the vehicle. All actuators are controlled by decentralized network nodes that are connected via a time-triggered communication system to facilitate safe control of the vehicle. This was the primary research purpose of the vehicle.

Volkswagen AG

Laumanns [2007] developed a flexible control system that generates optimal control performance from a given set of available actuators in a flexible vehicle during normal driving up to lateral accelerations of $4 \mathrm{m/s}^2$ on a high friction surface. The system was evaluated using an experimental vehicle that allows wheels to be braked individually, control the front steering angle, set an additional small steering angle at the rear axle, and perform active roll control.

BMW AG

A similar vehicle with active front-wheel steering, limited rear-wheel steering, brake control per wheel, and a central combustion engine is used by Krüger et al. [2010] to develop a control algorithm that can operate properly even if actuators fail. Such vehicles, which have all-wheel steering due to superimposed steering at the front wheels and limited rear-wheel steering (approx. 6 degrees), are becoming available in series vehicles, e.g., the BMW 7 series [Pfeffer and Harrer, 2011; Pruckner et al., 2012].

AUDI AG, Vehicle-in-the-Loop

In order to briefly consider research vehicles intended to investigate control systems that include environmental perception, the "Vehicle-in-the-Loop" by Bock [2008] will be described. The Vehicle-in-the-Loop equipment can be mounted on a standard series vehicle or prototype. The system includes additional measurement units for detailed analysis of the motion of the ego vehicle and detection of the driver's head position. Then, a simulation environment generates virtual traffic around the vehicle and provides the driver with an augmented reality view. Additionally, the electronic control system is provided with information about the virtual traffic as perceived by emulated sensors, e.g., Radar[5], camera, or Lidar[6] systems. As a result, the driver can evaluate vehicle handling during safety critical maneuvers, such as emergency braking, in a real vehicle, but can do so on a safe test ground within a (semi-) virtual environment. Moussa et al. [2012] follow a similar approach. Of course, the Vehicle-in-the-Loop equipment could be mounted on a highly flexible experimental vehicle to combine the advantages of the flexible platform and the (semi-) virtual environment.

This outlook concludes the list of introduced flexible experimental vehicles. As mentioned, this survey does not include vehicles that mainly focus on environmental perception, or demonstration/concept vehicles as built by all automotive manufacturers on a regular basis for automotive shows. If these vehicles are technically relevant to the research presented in this work and details are published, the vehicles are referenced in the appropriate sections.

3.2. Lessons Learned

Several conclusions about important technical features of flexible vehicles and the trends in their design over time can be drawn from the given projects. Technically, most of the vehicles are significantly over-actuated, meaning they feature more actuators than needed for "normal" driving operation. The vehicles mostly feature at least basic four-wheel steering, and a powerful system to individually control the brake torque at each wheel. Also, some projects extend the set-up by an active suspension system or a drive train that supports torque vectoring. Most vehicles feature drive motors that are sufficiently powerful for high performance tests. Based on these actuators especially the yaw rate, the side slip angle, and the pitch and roll movements of the vehicle can be controlled, which makes the vehicle a suitable test bed for various different control algorithms. Additionally, the dynamic behavior of a wide range of other reference vehicles can be emulated to compare and evaluate different vehicle set-ups during research and development. With regard to vehicle construction, the custom-built vehicles typically feature a "tube based" design, which allows an easy access to components, a high torsional stiffness, and a comparatively simple construction. Most projects do not include bodywork, as it is not needed for most tests and reduces accessibility. Still, special fields of application, such as the investigation of aerodynamics, require a hull. Depending on the research goals, flexible vehicles can reduce development costs and increase work efficiency during development when compared with project-specific prototypes.

Construction and actuators

When the trends in design and application of the prototypes are considered, a change in research focus becomes obvious. Starting from prototypes to investigate the mechanical design and axle geometry, newer vehicles are increasingly targeting electronics and sophisticated control of the available active components of the drive

Trends in the design

[5]Radio Detection and Ranging
[6]Light Detection and Ranging

and suspension system. This development may have be caused by a couple of influences. Certainly, the limited computational power and limitations in the available active components in earlier projects led to the designs mainly based on mechanics. This becomes especially obvious when the difficulties in online control of the actuators in the early outlined projects are taken into account. Additionally, modern vehicles already feature mature, high performance suspension systems specifically designed for a given type of vehicle. These systems evolve stepwise by fine-tuning and require high precision measurements for the intended target vehicle. For these experiments, flexible vehicles may be unsuitable due to the remaining deviations from the final target platform. In comparison, active components in the suspension system or electric drives that are increasingly becoming available in series vehicles still offer a huge potential to improve the driving performance and generate noticeable benefits for the driver. Consequently, it seems reasonable for the research focus to shift accordingly. Nevertheless, mechanically modular and easily modifiable vehicles have gained a new right to exist: the components of series vehicles are usually subject to intellectual property restrictions, which makes it hard to implement novel algorithms in research. Moreover, the novel mechatronic components are hard to integrate into series vehicles both electronically and in terms of the vehicle package. Thus, a flexible vehicle can serve as a "carrier" for components, and universities can profit from these easily accessible and comprehensible vehicles for teaching.

Conclusion

In general, the flexible experimental vehicles have proven suitable for a wide range of applications. The MOBILE project will stick to some of the outlined trends to derive a suitable development platform. Especially, three aspects of the introduced trends carry through to the developed vehicle:

- High performance actuators are installed in the vehicle to provide the basis for novel control approaches and to allow simulating the behavior of a wide range of less powerful systems.
- The mechanical design shall facility easy exchange of components for research purposes without restrictions by intellectual property of companies.
- The mechanical flexibility in terms of reconfigurability of the vehicle dynamics is restricted to an extent that suffices to achieve similarity to series vehicles for research purposes. The main focus is put on the flexible electronics in the vehicle to develop novel control algorithms.

The resulting design decisions and requirements for the vehicle and the associated toolchain will be introduced in the next chapter. Cha. 9 will add a simplified and adapted safety analysis of the experimental vehicle, which has so far rarely been done in other projects.

4 Structure and Elements of the Toolchain

"Man is a tool-using animal. Without tools he is nothing, with tools he is all."

Thomas Carlyle

To efficiently make use of the flexible experimental vehicles for iterative development, appropriate support by tools is required. For the tool design, challenges arise in multiple ways. A cost efficient tool has to ensure sufficient functional safety during testing while being easy-to-use for the developer and easily extensible. Especially, the functional safety requires close coupling of the tool design with the set-up of the experimental vehicle.

Structure of the toolchain

In detail, the toolchain designed in this work supports the developer during evaluation of the first implementation of a novel algorithm in simulation and migration of the resulting system onto an experimental vehicle for testing under realistic driving conditions. A 1:5 scaled vehicle complements the toolchain, especially by supporting the stepwise safe development of safety critical applications as summarized in Fig. 4.1. Extending the main goal to support the developer, the toolchain is subject to requirements resulting from the intended application in an evolving research environment:

- The toolchain itself has to be easy to modify, as new applications or hardware components may require adaptation or extension of the toolchain.
- Compatibility with well-known development tools in the field of vehicle control, especially Matlab/Simulink, has to be achieved.
- Commercial tools with sufficient access to the underlying processes are preferable to custom written tools in order to limit the effort for tool maintenance.
- In terms of execution platforms, both onboard computers and cost-efficient microcontrollers should be usable with the toolchain.

The following sections outline the most important components of the toolchain to both introduce the working basis for experiments carried out in this thesis and derive special requirements for the experimental vehicles due to their usage as a part of the toolchain.

P.J. Bergmiller, *Towards Functional Safety in Drive-by-Wire Vehicles*,
DOI 10.1007/978-3-319-17485-3_4

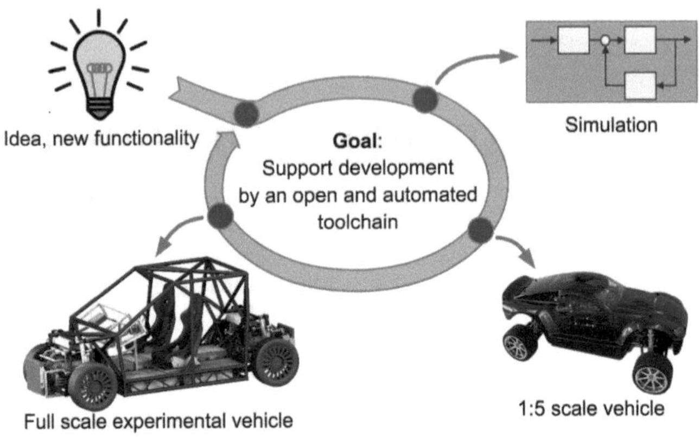

Figure 4.1.: Goal and basic structure of the toolchain developed in this work

4.1. Real Vehicles to Assist Development

The need for practical tests in the research project MOBILE is addressed by flexible experimental vehicles. These vehicles allow novel control systems to be developed to a stage of completion that suffices to verify practical applicability. For that purpose, a scaled and a full-scale vehicle were constructed as part of this thesis. Four core requirements given in Tab. 4.1 and three main constraints (Tab. 4.2) guided the development. The requirements are grouped into such that ensure the flexibility of the vehicle and others that target sufficiently safe testing. Thus, the experimental vehicles have to bridge the gap between good usability in various projects and functional safety at limited costs. The following sections outline the basic set-up of the vehicles. More details on the vehicle electronics and the algorithms executed on the vehicles will be given in Cha. 5, 7, and 8.

4.1.1. Experimental Vehicle MOBILE[1]

The full-scale vehicle MOBILE was custom built by the Institute of Control Engineering and the Institute of Engineering Design at TU Braunschweig. The basic actuator set-up of the vehicle was derived from the use cases and the resulting requirements (Tab. 4.1). The requirements for flexibility dominated → (R1) and led to a design as a full electric vehicle with by-Wire control for the propulsion, braking, and steering system.

Electric drive

The electric drive concept of MOBILE contributes to a powerful base configuration by ensuring flexibility in control of the longitudinal dynamics. The research project InDrive demonstrated that the longitudinal behavior of a target vehicle

[1]Parts of this section have been pre-published by the author in Bergmiller [2013].

Table 4.1.: Requirements for the experimental vehicles

Providing a flexible experimental vehicle	
R1	*Mechanical and electronic modularity:* The vehicle has to feature mechanical and electronic modularity to allow the easy exchange of components for research and testing. Nevertheless, the base configuration of the vehicle has to be sufficiently powerful to already support a wide range of driving experiments.
R2	*Open-source vehicle:* The software executed by the electronic control units in the vehicle has to be easily accessible and exchangeable. The vehicle can be seen as an "open-source vehicle" that is readily available for research tasks. Compatibility with Mathworks' graphical programming environment Simulink is desired to shorten training periods.
Guaranteeing sufficiently safe testing	
R3	*Functional safety:* Although the experimental vehicle is only operated on test tracks, the vehicle should fulfill basic safety requirements and tolerate one independent fault[a] with a sufficiently high probability.
R4	*Limited degree of hardware redundancies:* The degree of hardware redundancies for safety purposes shall be reduced. In turn, the safety concept shall exploit functional redundancies between different types of actuators that are available in the vehicle.

[a] A fault is an "abnormal condition that can cause an element or item to fail" [ISO 26262-1, 2011, p. 7].

can be simulated by a powerful experimental vehicle given low latencies in traction control [Cornelsen et al., 2011]. The electric drive concept of MOBILE with a peak power of about 100kW at each of the four wheels and low latencies in torque control can fulfill these requirements. Additionally, the independent drive of each wheel allows a yaw control via torque vectoring. The benefits and risks of torque vectoring with regard to vehicle control are, e.g., evaluated by Euchler et al. [2010], Piyabongkarn et al. [2007], or Rohe [2012]. Mechanically, the electric components support the modularity by facilitating the easy exchange of the drive units.

Four-wheel steering

Four-wheel steering adds further flexibility to the vehicle control system. In general, the steering system can be implemented as a rack actuating type, a tie-rod actuating type, or a knuckle actuating type [Park et al., 2005]. In order to be able to individually steer each wheel and based on the components available on the market, the tie-rod actuating type was implemented. Thus, different steering geometries and steering concepts can be emulated by simple software modifications. In terms of performance, Wilwert et al. [2005] consider a ±40 degree steering angle per wheel and a steering rate of approx. 40 degrees per second as desirable (compare

Table 4.2.: List of constraints

C1	The MOBILE project is a "university only" project and thus has to rely on students writing their theses on individual development tasks. The resulting multiple small work packages have to be parallelized and coordinated.
C2	All tasks worked on by the project partners have to stem from the associated research fields. Other parts have to be sourced externally, e.g., actuators.
C3	The project is subject to strict financial limits. Thus, mostly "off-the-shelf" components have to be relied on.

Heiner and Thurner [1998] for an OEM's view). This steering angle approximates typical characteristics of front-wheel steering systems in series vehicles with front-wheel drive [Pfeffer and Harrer, 2011]. For MOBILE, each individual steering system features an adjusting range of approx. ±43 degrees and a steering rate of 130 degrees per second at nominal load. Thus, also highly dynamic maneuvers are possible.

Electro-mechanical braking

The electro-mechanical braking system for MOBILE was designed by Vienna Engineering and is supposed to outperform most hydraulic brake systems in terms of reaction times. This makes it possible to control slip precisely and to analyze seamless integration of recuperative and mechanical braking to increase energy efficiency [Pruckner et al., 2012]. Additionally, the braking system renders hydraulic components in the vehicle unnecessary and thus little impacts vehicle package. The electromechanical system is designed to ensure a 1g deceleration of MOBILE at a maximal weight of 2.1 tons including passengers. First tests on a test bench indicate that these requirements are outperformed. The safe state of the individual brakes is defined as a state without any brake torque as also demanded in the literature [Johannessen et al., 2004a; Sinha, 2011].

Power supply

MOBILE is powered by two independent lead-acid battery packs providing approximately 300V each. In future, these batteries are planned to be exchanged by lithium ion batteries with a pack voltage of 400V and higher energy density. Additionally, two independent low voltage circuits at 48V and 12V are associated with each high voltage battery and supply actuators and vehicle electronics (48V for steering, 12V for braking and other electronics, → C3). The parallel structure of the power supply system reduces the overall failure rate → R3 and limits the maximal currents drawn from the main battery packs at peak load. Also, steering and braking actuators at diagonal positions in the vehicle are connected to a common power supply, which results in less disturbance of the vehicle handling if one supply fails. This corresponds to the braking system design in series vehicles (ECE R13[2]) and is frequently replicated for Brake-by-Wire systems proposed in the literature [Rieth, 2012; Papadopoulos et al., 2001]. The powerless steering actuators

[2]United Nations Economic Commission for Europe: Brake System Homologation, see appendix Sec. A.1

are back-drivable and thus can be moved by torque at the wheels applied by the drive motors given a suitable axle geometry [Dominguez-Garcia et al., 2004].

User interface

The by-Wire architecture makes it possible to design the user interface flexibly. All input devices can be exchanged on demand. In a base configuration, a force-feedback steering wheel, a force-feedback brake pedal, and a gas pedal are available to the driver. These units can provide feedback on the road surface and the current driving condition. A flexible touch-screen based visualization that will be briefly introduced in Sec. 4.1.3 allows easy access to all measurements available in the vehicle. All actuators including the ones of the user interface are controlled by a network of specifically developed ECUs featuring a wide range of bus interfaces and digital and analog inputs and outputs.

Differentiation of MOBILE

To conclude, Fig. 4.2 provides an overview of the important components of MOBILE. Further details can be found in Bergmiller and Maurer [2012]. If this set-up is compared with the lessons learned from the state-of-the-art on experimental vehicles given in Sec. 3.1, it becomes obvious that MOBILE follows the trends in the design of experimental vehicles. MOBILE features an actuator set-up that allows to control steering, braking, and torques for each wheel individually. As pointed out in the state-of-the-art, this supports a wide range of experiments. In terms of the actuators, MOBILE is only missing an active suspension system, but the vehicle package was designed such that, if needed, adding such a system should be possible. In combination, the modular set-up, the flexible and powerful electrics and electronics, and the frame-based design provide a unique opportunity to investigate mechatronic components in detail under realistic driving conditions. Mechanical modifications to MOBILE are easily possible although, e.g., special mechanisms to quickly modify the suspension set-up "on the test site" as in some of the introduced vehicles are not available, as these mechanical features are not focused. The vehicles closest related to MOBILE are the "X1" vehicle from Stanford and the "In-Vehicle Simulator" by Nissan. Especially, the latter vehicle has some disadvantages compared to MOBILE due to the restrictions of the actuators and control electronics available at the time of construction of the vehicle. In summary, the chosen actuator set-up for MOBILE ensures flexibility for research on mechatronic components and software. The vehicle provides a powerful base set-up with a high degree of functional redundancy between different types of actuators, which can barely be found in the other research vehicles.

4.1.2. Modular Adaptable X-by-Wire Vehicle (MAX)

The 1:5 scaled vehicle "MAX" (Modular Adaptable X-by-Wire Vehicle) extends the toolchain to reduce the gap between simulation experiments and the experiments on the full-scale vehicle MOBILE. For that purpose, scaled models have long been relied on in different fields of research: Brennan and Alleyne [2001b], Hilgert [2005], Johannessen et al. [2004a], König et al. [2006], and Verma et al. [2008] rely on scaled vehicles to design mechatronic components in the automotive field but also reference applications in other fields, such as aerospace or marine. Hilgert [2005]

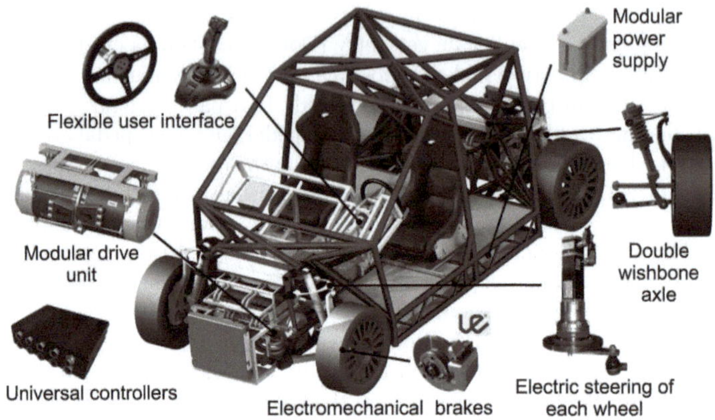

Figure 4.2.: Actuators and mechanical set-up of MOBILE

summarizes the two main benefits of scaled vehicles for development. (a) The scaled models can significantly reduce costs compared to the construction (and operation) of early full-scale prototypes if simulation cannot provide a sufficient accuracy; and (b) the scaled models allow safe evaluation of safety critical applications. Both aspects are exploited by the presented toolchain, but especially the chance to safely pre-develop safety critical control algorithms on the scaled vehicle stands out.

Pi-Theorem

With some restrictions, the results obtained with the scaled model can be transferred to the full-scale vehicle if transformed according to the Buckingham Pi-Theorem, which is detailed, e.g., by Yarin [2012] and explained for the use case in this thesis in the appendix (Sec. A.2). Verma et al. [2008] rely on such a scaling strategy for the investigation of the longitudinal dynamics of a vehicle including a detailed drive train model, and Hilgert [2005] targets the lateral dynamics of a scaled vehicle to evaluate path planning capabilities. Brennan and Alleyne [2001a,b] analyze the suitability of scale vehicles for controller development and investigate the use of non-dimensional vehicle dynamics for robust scalable vehicle control. Table 4.3 outlines the important scaling factors for MAX as derived in this work. The scaled-up values are compared with MOBILE. The table assumes a reference scaling factor of 1:4.86 rather than 1:5 derived from the wheelbase. The obtained results conform with similar investigations made by König et al. [2006], who developed a steering controller based on a scaled vehicle. The scaling rules generating the factors given in Tab. 4.3 are given in the appendix (Sec. A.2).

Restrictions

One of the main limitations of scaled vehicles observed in the MOBILE project is that the driver is not onboard the vehicle and does not really feel the vehicle handling. Also, it is important to note that for quantitative validity of the obtained data, the scaled vehicle has to undergo detailed investigations, comparable to full-scale vehicles. Especially, the tire road contact has to be considered. There-

Table 4.3.: Comparing properties and measurements across different sized vehicles

Properties/measurements for vehicles	MAX	MAX scaled	MOBILE	Scaling factor
Track width (m)	0.4	1.94	1.75	4.86
Wheel base (m)	0.56	2.72	2.72	4.86
Mass (kg)	16	2000	2000	125
Max speed (m/s)	11[a]	53	44[b]	4.86
Wheel diameter (m)	0.14	0.68	0.64[c]	4.86
Yaw rate (rad/s)	$\dot{\psi}$	$\dot{\psi}$	$\dot{\psi}$	1
Side slip angle (rad)	β	β	β	1
Max accelerations (m/s^2)	5[d]	24.3	10[e]	4.86
Lateral stiffness (N/rad)	250[f]	151875	-	607.5
Moment of inertia ($kg{\cdot}m^2$)[g]	0.44	1299	-	2952, 45

[a]Highest measurable speed on the test track; vehicle can go faster
[b]Depends on the diameter of the mounted wheels
[c]Maximal acceptable diameter for the implemented mechanical set-up
[d]Mainly limited by the surface parameters
[e]Limited by tires and surface
[f]Values ranging from 230N/rad (front axle) to 285N/rad (rear axle) were identified by matching a virtual model and measurements from a test drive. Hilgert [2005] identifies 762N/rad for a similar scaled model on a high friction surface with roughly double the friction coefficient.
[g]Around vertical axis

fore, Hilgert [2005] tests the tires of the scaled vehicle at a tire test rig. Still, full quantitative transferability, especially for highly dynamic maneuvers under varying surface conditions, remains questionable, as, e.g., also the measurement errors are scaled up. In the MOBILE project, the precise transfer of measurement results is not focused on. Mainly, the qualitative operation and basic validity of the algorithms are evaluated using the model vehicle MAX.

Set-up of MAX

Fig. 4.3 provides an overview of the actuator and electronics set-up of MAX. MAX features independent four-wheel steering and one drive motor per axle. The actuators are controlled by up to four microcontroller based electronic control units and an onboard computer that also provides wireless access to the vehicle. The electronic control units used in the scaled vehicle feature the same microcontrollers as the units used in MOBILE. All controllers communicate via a FlexRay bus with redundant data transmission on the two available channels. MAX is equipped with wheel speed sensors, accelerometers, gyroscopes, GPS, and a perambulator wheel to measure all important dynamic state variables including the side slip angle. Compared to MOBILE, the scaled vehicle is simplified in several aspects with functional relevance:

- The wheels of an axle cannot be driven individually.

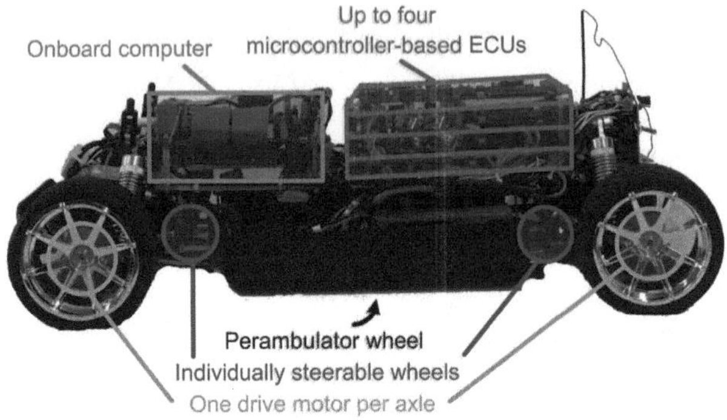

Figure 4.3.: Components of the modular experimental vehicle (MAX)

- The wheels can only be braked with the drive motors.
- Only a reduced set of controllers (up to four) and an onboard computer with reduced computational power are available.
- The sensors are not available redundantly.
- The user inputs are provided via remote control.

Relating this set-up with the requirements for the experimental vehicles (Tab. 4.1), it becomes obvious that the requirements for the mechanical and electronic modularity → R1 and on easy accessibility of the source code → R2 are met. The transfer of software implemented for the scaled vehicle onto the full-scale vehicle is supported by the identical microcontrollers used in both vehicles and by the tool environment briefly outlined in the following section. Requirements for functional safety given in Tab. 4.1 are fulfilled due to the scaled size of the vehicle and the safe testing environment.

4.1.3. Assisting Tools

To assist research based on the experimental vehicles, a software toolchain was set up. Four main goals derived from the top-level requirements → R2 guided the implementation:

- Support the "open-source" concept on multiple platforms (microcontrollers and onboard computers).
- Enable easy code transfer between simulation and experimental vehicles.
- Relieve the developer of low-level and repetitive implementation tasks by automatic code generation from graphical models.
- Rely on commercially available tools if possible.

To achieve these goals, an approach based on Matlab/Simulink was chosen. The user implements new applications in C code or graphically in Simulink. Then, the graphical components are converted into C code by Simulink's "Embedded Coder". The resulting code is taken as an input for the parts of the toolchain that were implemented in this work. It is enriched by appropriate drivers for the target platform and embedded into a custom written mini operating system with small resource consumption that fulfills timing requirements, allows basic task scheduling, automatic synchronization with the global clock in the vehicle, and provides basic diagnostics for onboard peripherals. Two different operating systems are available for the controllers, which will become important when the fault tolerance of the system is investigated. To configure the hardware platform and the operating system, suitable Simulink blocks are provided to the user. Also, Simulink blocks that represent available signals on the FlexRay bus are automatically generated from a FIBEX[3] file. In summary, the developer can implement a novel application graphically and either execute the application in simulation, flash the application on any microcontroller on one of the experimental vehicles, or run it on an onboard computer.

Code generation

The advantages of such an automated approach to code generation have already been investigated in several projects and have proven useful during development of highly safety critical applications for series vehicles, such as BMW's active steering system [Zoelch et al., 2006]. Especially, the graphical programming approach supports modularization and reduction of programming errors. Nevertheless, no fully integrated toolchain that supported both novel bus systems, such as FlexRay, highly time critical application execution, and multiple execution platforms (different microcontrollers and simulation computers) was available on the market when the toolchain outlined in this section was implemented. Reconsidering the implemented mini operating system from 2008, AUTOSAR 4.0 may become an option to replace the existing system. The reworked standard now supports FlexRay sufficiently and increasingly supports highly time and safety critical applications, which was not the case in earlier versions [AUTOSAR, 2010, 2012]. Still, a full or partial shift towards AUTOSAR will have to be evaluated depending on the assumed benefits for future research projects. So far, the existing toolchain has been used in the MOBILE project but also in related projects at the Institute of Control Engineering as the already referenced InDrive project [Cornelsen et al., 2011]. More details on the toolchain can be found in Bergmiller [2008] and Bliedung [2010].

Evaluation

AUTOSAR

For visualization of measurements and monitoring of the vehicle state, a graphical user interface is provided [Bergmiller et al., 2011a]. The user interface allows flexible visualization of all data available on the FlexRay backbone bus based on different visualization elements, such as gauges, graphs, or numerical displays. These units can be freely added, formatted, and positioned to generate a user specific dashboard layout. Technically, a server application receives data from the FlexRay bus and streams it to multiple clients that are connected via LAN or Wireless LAN allowing

Monitoring and visualization

[3]Field Bus Exchange Format [Tsitlakidis et al., 2008]

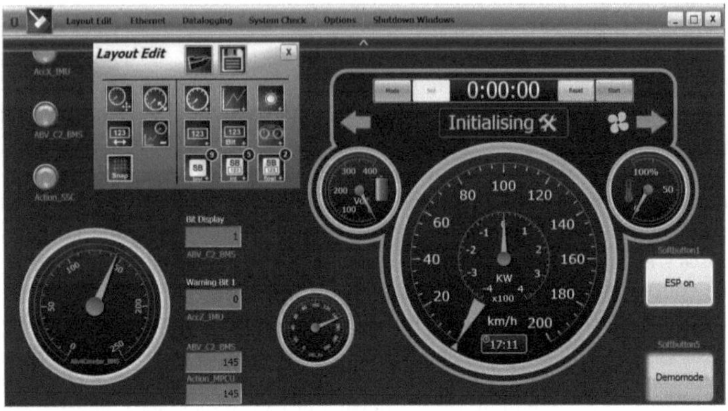

Figure 4.4.: Screenshot of the flexible user interface

access to vehicle data both onboard and off-board. Multiple screens per client and touch based operation are supported. Additionally, authorized users can add input elements, such as buttons or numerics, to the user interface. These inputs are then transmitted via FlexRay and received by the network nodes. This allows the user to parametrize or trigger applications while driving. Figure 4.4 visualizes an example set-up of the graphical user interface. The system is described in more detail by Güntner [2012], Ibele [2009], and Laskowski [2011].

4.2. Simulation Environment for Development and Testing

An additional simulation environment based on Matlab/Simulink allows the testing of novel control algorithms before the migration onto the real vehicles. The developed system is structured modularly and can easily be extended or modified. The code of all simulated components is fully accessible to the developers, and the sufficient accuracy of the simulation system for the use cases in this thesis has been shown by comparing simulation results with measurements from the 1:5 scaled vehicle. In future, further benchmarks will have to be created for the full-scale prototype. Figure 4.5 summarizes the available software components that are introduced in the following.

4.2.1. Vehicle Models

The virtual vehicles implemented for closed loop simulation experiments focus aspects relevant to vehicle dynamics. All models are available as "drag-and-drop"

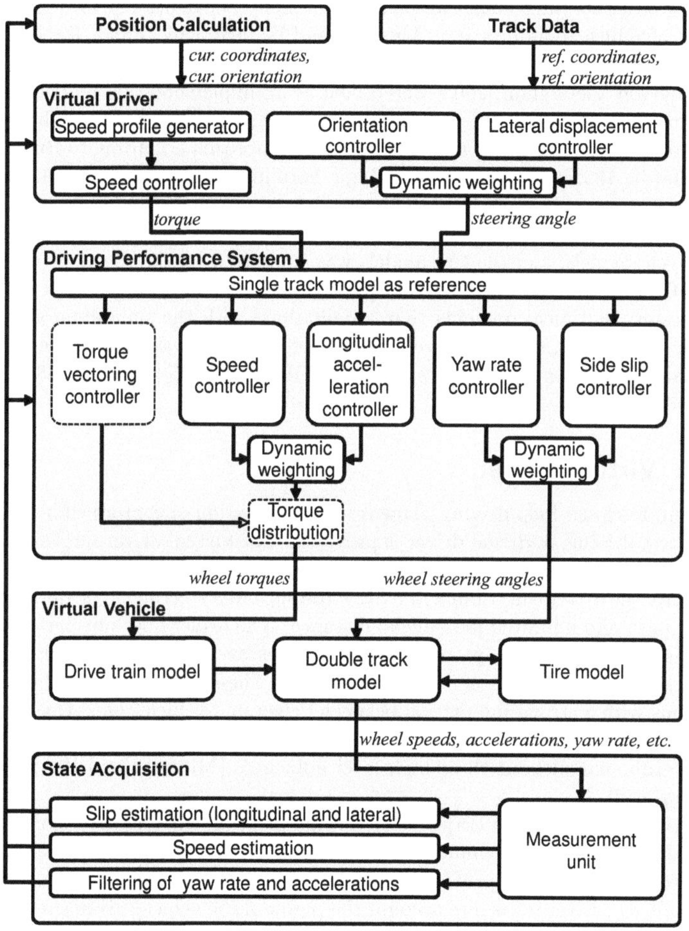

Figure 4.5.: Software components for simulation or online use; control modules in dotted boxes were only tested in simulation

Simulink blocks and can be executed in real-time on microcontrollers or an on-board computer if needed.

Double track vehicle model

As given in Fig. 4.5, the central vehicle model is a non-linear double track model that is supplemented by a drive train model and a tire model. The associated formulas to model the vehicle can be found in many books dealing with the fundamentals of vehicle dynamics, e.g., Mitschke and Wallentowitz [2004]. For this work, the detailed deduction and description of a double track model and a drive train model by von Vietinghoff [2008] were taken as an implementation guideline. As a main simplification, the used model does not cover suspension system influences on roll- and pitch dynamics. To model the tires, an approach according to Burckhardt [Burckhardt, 1993, pp. 24] and the Magic Formula tire model [Pacejka, 2012, p. 165] have been implemented. Both tire models are parametrized by experimental data. The implementation and verification of the double track model, the simplified bicycle models, and the tire models was supported by the theses of Gemeiner [2011], Goldschmidt [2012], Lieberam [2011], Schwarz [2012], and Töpler [2010].

All results and measurements that are obtained with the simulation system or with the experimental vehicles are given in a coordinate system according to DIN 70000 and the upcoming replacement norm DIN ISO 8855 [DIN ISO 8855, 2011; DIN 70000, 1994].

4.2.2. Virtual Driver

To facilitate closed-loop driving maneuvers in simulation or reproducible test runs during real driving, a virtual driver is provided. The virtual driver implemented in this thesis is intended to keep the (virtual) vehicle on a given reference trajectory. Therefore, the driver can command a drive torque and a steering angle derived from a given path and a limited look-ahead distance. The driver does not perform path planning, dynamically adapt to different vehicles as humans do, or emulate the human perception and action derivation process. These aspects have to be covered if systems with a strong interaction between driver and vehicle control systems are investigated [Asano et al., 1991]. Still, the basic operation of the control systems can be evaluated without these high level influences [Mitschke and Wallentowitz, 2004, pp. 644].

Approach

To accomplish the given trajectory following task, the virtual driver controls the longitudinal and lateral dynamics of the vehicle in three steps. (1) To start with, the driver derives a speed profile for a configurable look-ahead distance, e.g., 100m. Therefore, the driver takes into account the power of the vehicle, given speed limits, and tire saturation. Tire saturation at a given point on the race track is estimated based on the assumed friction coefficient and the expected speed and acceleration at that point. (2) Then, a PID[4] speed controller commands drive torques to follow the generated speed profile. (3) Lateral control of the vehicle is performed based on an approach similar to the one introduced by Sharp et al. [2010], which uses both the lateral displacement and the alignment of the vehicle. Fig. 4.6 outlines

[4]proportional-integral-derivative

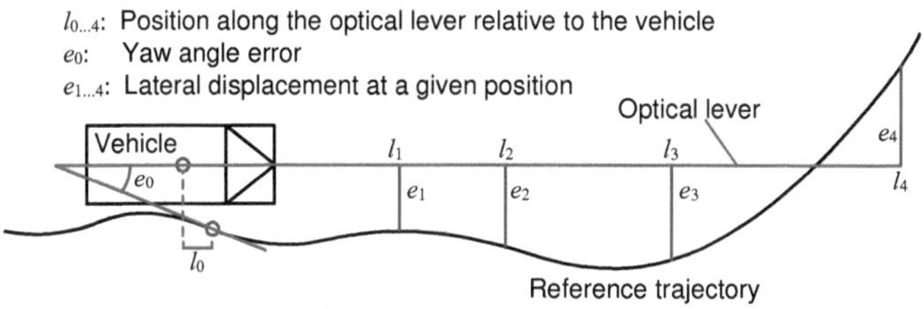

Figure 4.6.: Inputs to the virtual driver for lateral control as adapted from Sharp et al. [2010]

these two inputs. The lateral displacement is considered at several points along an "optical lever" as introduced by Sharp et al. [2010]. The results of the controllers for each of the positions along the optical lever are weighted, saturated, and summed to generate the steering angle output from the lateral displacement. Additionally, the current yaw angle error is considered by Sharp et al. [2010] and added to the weighted sum generated from the lateral displacement values. As can be seen in Fig. 4.6, the driver implemented in the MOBILE project takes the reference angle from a point slightly ahead along the track. Thus, the virtual driver can be designed to cut corners if needed. Also, deviating from the approach by Sharp et al. [2010], the outputs of the two controllers based on angle deviation and lateral displacement are weighted depending on the deviation from the reference track. If the vehicle is close to the reference track, control focuses on angle error; otherwise it starts to favor the lateral displacement controller. The introduced virtual driver is fully accessible to the developer and can easily be adapted. Especially, the parameters to adjust the planned speed profile, such as the assumed friction coefficient or the acceptable lateral displacement, hugely impact the way the driver follows the given trajectory. For example, to test the performance of a stability control system, the assumed friction coefficient can be set higher than the available friction coefficient, which then leads to a "risky" driving style.

Tuning the driver

The virtual driver represents a useful extension of the presented simulation environment and has performed well for the simulation experiments. It was also ported onto MOBILE and successfully applied to follow a given trajectory. Future work can gradually extend the virtual driver and, e.g., evaluate alternative control strategies based on the introduced control variables. The current basic PID control is a first step.

Evaluation

4.2.3. State Acquisition and Estimation

The state acquisition unit collects sensor data throughout the vehicle and derives the current speed, the side slip angle at the center of gravity, and the filtered yaw rate and acceleration signals. In full simulation mode, the state acquisition unit can be replaced by a data forwarding system, as all important values are directly available from the double track model. For a more realistic investigation and the usage in real vehicles, appropriate estimation and filter modules are available.

Side slip

During dynamic driving, high side slip angles can result. The proper estimation or measurement of this angle is vital for a multitude of stability control applications. For MOBILE, a simple side slip estimation based on a linear bicycle model with a speed dependent parametrization and a Luenberger observer [Luenberger, 1979] is provided. During the driving experiments with the scaled vehicle, the side slip was measured with a perambulator wheel.

Speed

In an all-wheel drive vehicle, the real speed becomes hard to estimate especially if high torques are applied to all wheels. Without additional equipment, a direct measurement of the vehicle speed is no longer possible, as all four wheels may feature significant slip. In this work, the approach proposed by Kobayashi et al. [1995] is followed. Two Kalman filters pre-filter the measurements in the vehicle. One filter targets the measured accelerations, the other the measured wheel speeds. Both signals are assumed to be Gaussian distributed. A third Kalman filter takes the pre-filtered acceleration and speed outputs as inputs and generates a merged speed signal. The covariance matrix entries of the system and measurement noise associated with the two inputs are weighted by a Fuzzy Logic that takes the current driving situation into account. For example, during hard acceleration phases, trust is shifted from the wheel speeds to the acceleration measurements. A similar fuzzy based approach is presented by Hilgert [2005] but without combining it with Kalman filters as introduced by Kobayashi et al. [1995]. The implemented speed estimation generated good results during benchmark scenarios [Lieberam, 2011].

4.2.4. Track Generator

To generate the reference track data for simulation experiments, a simple track generator was implemented, which allows to generate a random track manually from basic elements: curves with constant radii, clothoids, straight segments, and start and stop elements. The developer can combine these elements and modify parameters as length, radius, and speed limit for each element to generate an overall track.

4.2.5. Driving Performance System

The Driving Performance System concludes the list of implemented modules. It is intended to integrate several available controllers to support the driver during the driving task. Currently, only some basic control algorithms are available, which will be outlined in Sec. 7.2. These algorithms were implemented and evaluated on the 1:5 scaled vehicle but are compatible with the existing simulation environment.

Part III: A Novel EE Architecture for Drive-by-Wire

This part introduces a novel EE architecture for use in flexible Drive-by-Wire vehicles as MOBILE. In combination with special mechanisms, the architecture assures functional safety while limiting required hardware redundancies and facilitating evaluation of novel, little tested control systems.

5

The EE Architecture of the Experimental Vehicle MOBILE[1]

"To create architecture is to put in order.
Put what in order? Function and objects."

Le Corbusier

The vehicle MOBILE is a vital part of this thesis in a twofold way: It contributes to the outlined toolchain, and it is itself a demonstrator for novel structures and mechanisms to achieve both sufficient functional safety and flexibility in Drive-by-Wire vehicles. This chapter details the EE architecture[2] of MOBILE and focuses on the part of the onboard network that directly impacts by-Wire control and thus features the highest safety criticality.

Flexibility vs. safety

The development of the EE architecture is driven by the core requirements for the vehicle introduced in Sec. 4.1 (Tab. 4.1). The unique combination demanding modularity, flexibility, and limited costs on the one side and guaranteed controllability on the other side represents the main challenge. The architecture proposed in this section demonstrates novel approaches to efficiently address these matters in a Drive-by-Wire vehicle.

Structured architecture derivation

The EE architecture is introduced stepwise at different hierarchical layers and from different views[3]. Figure. 5.1 outlines the resulting steps. The first two steps have already been covered in the previous chapters. There, the top-level functional goals and constraints in terms of the intended application (step 1 in Fig. 5.1) guided the definition of the basic mechanical and actuator set-up of the vehicle (step 2 in Fig. 5.1). The following architecture derivation takes these results as an input, and

[1]Parts of this chapter have been pre-published by the author in Bergmiller [2013].

[2]ISO/IEC/IEEE 42010 [2011] defines architecture as "fundamental concepts or properties of a system in its environment embodied in its elements, relationships, and in the principles of its design and evolution" [ISO/IEC/IEEE 42010, 2011, p.2].

[3]The architecture of a system can be described from different views depending on the goal of the description. Examples are business, process, functional, or hardware views [Masak, 2010]. For some of these views, guidelines for standardized diagrams exist, e.g., UML for software systems [Starke, 2008]. ISO/IEC/IEEE 42010 [2011] defines architecture view as "work product expressing the architecture of a system from the perspective of specific system concerns" [ISO/IEC/IEEE 42010, 2011, p. 2].

P.J. Bergmiller, *Towards Functional Safety in Drive-by-Wire Vehicles*,
DOI 10.1007/978-3-319-17485-3_5

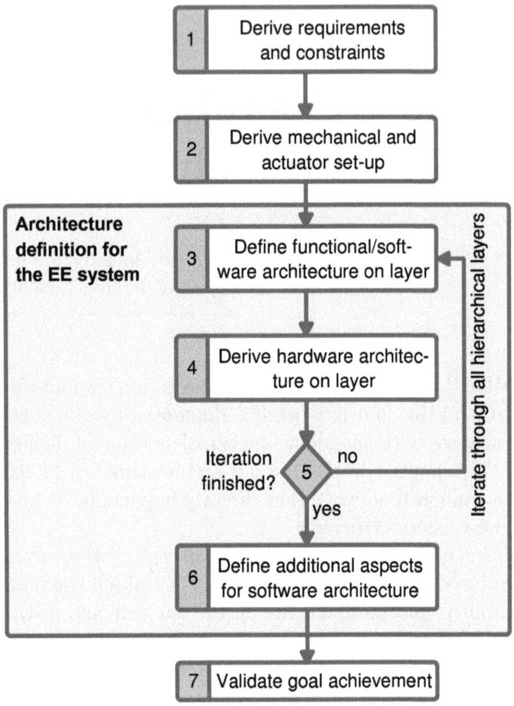

Figure 5.1.: The architecture definition process for MOBILE

includes additional factors that extend the project specific requirements. As a basis for that, Maier and Rechtin [2009] distinguish "normative (solution based)", "rational (method based)", "participative (stakeholder based)", and "heuristic (lessons-learned)" [Maier and Rechtin, 2009, p. 1] methodologies. To cover important normative and heuristic influences, Sec. 5.1 provides a summarized state-of-the-art on structures of Drive-by-Wire systems, used mechanisms, and best-practices. Cha. 6 will introduce a novel method to analyze the system safety that was used both during the system design and evaluation (rational methodology). Participative aspects are not detailed in this contribution, as the developers and users of MOBILE will be identical in the first step. Possible future use cases of MOBILE are taken into account by the flexible design. Based on the mentioned inputs and requirements, the functional/software and hardware architecture of the EE system is defined at all hierarchical layers (steps 3 to 5 in Fig. 5.1). Figure. 5.2 introduces the defined hierarchical layers[4]. The functional and software architecture partially merge at

[4]In the following, the hierarchical layers are always referred to as "X level" or "X layer" (with

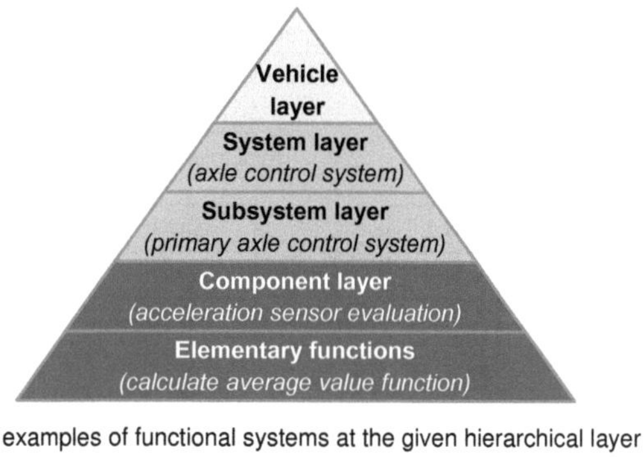

Figure 5.2.: The hierarchical layers for the architecture definition

higher hierarchical layers. Thus, no dedicated software views will be presented at the layers, but some special aspects are covered in step 6 (Fig. 5.1). Finally, the overall system is evaluated with regard to the goal achievement in step 7 (Fig. 5.1).

5.1. State-of-the-Art: EE Systems of Drive-by-Wire Vehicles

This section provides an overview of the existing approaches to design Drive-by-Wire systems but only covers fully electronic systems. Solutions that include a mechanical/hydraulic fall-back layer are not detailed, such as the systems by Emery [1998], Zuo et al. [2005], or the recently revealed Steer-by-Wire system by Nissan [Heise, 2012; Ramey, 2012]. Also, designs with only one central controller as in the vehicle by Park et al. [2005] or several other similar research vehicles that were built without considering functional safety are neglected. A centralized architecture would require huge wiring effort to achieve functional safety, have to deal with EMI[5] challenges, and features little modularity. To organize the information presented in this section, Fig. 5.3 proposes a generic view of a by-Wire system. The numbers in the figure reference key aspects, for which the state-of-the-art will be presented.

System structure ❶

Most by-Wire systems for vehicles investigated in research, e.g., by Armbruster [2009], Heiner and Thurner [1998], Sinha [2011], Wilwert et al. [2005] or Zuo et al. [2005], split the system into two physically separated sections as generalized in Fig. 5.3. One section contains the user interface, the other section controls the actuators mounted at the axle, e.g., the steering actuators, brake actuators, or

X standing for vehicle, system, etc.) to clearly distinguish between referencing a specific layer and the general use of the words system, component, or element.

[5]Electromagnetic interference

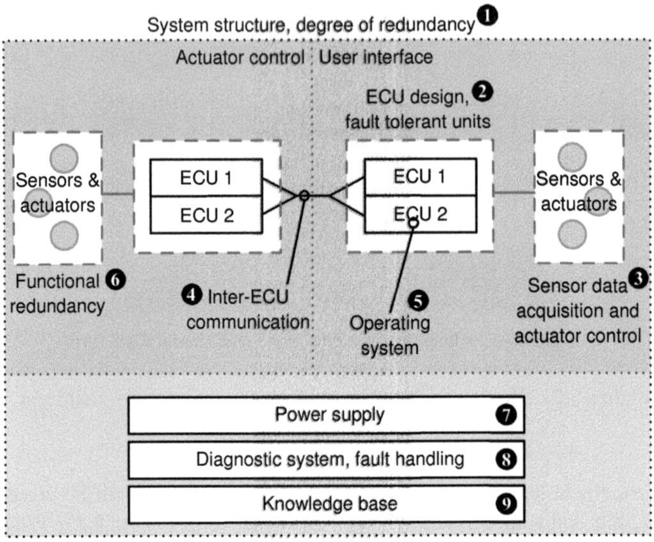

Figure 5.3.: A generic by-Wire system to structure the state-of-the-art review

other actuators as for control of the vertical dynamics. To capture user inputs, different types of devices are investigated that can also include actuators to provide feedback on the road surface and the driving situation to the driver [Winner and Heuss, 2005].

Redundant local controllers ❷

The actuators and sensors are controlled and monitored by closely mounted decentralized ECUs[6]. Each of these safety critical ECUs is usually available redundantly, as no single unit achieves the required failure rates. Still, the degree of hardware redundancy should be kept as low as possible due to the production costs. In general, two components for one task in combination with a sufficiently powerful diagnostic and decision unit and a fail-silent behavior of each component are assumed to be able to achieve the required failure rates [Kirrmann and Großpietsch, 2002; Miller, 2007; Wilwert et al., 2005][7]. Several systems featuring this structure can be found in the literature [Armbruster, 2009; Hasan and Anwar, 2008; Sinha, 2011; Wilwert et al., 2005]. Such a combination of two (or more) units is then regarded as a fault tolerant unit[8]. If all redundant units are permanently turned on, the system is denoted as operating in hot-standby mode [Neudörfer, 2011]. This shortens the take-over times in case of failure of the primary unit. In

[6]Electronic Control Unit

[7]Note: For Fly-by-Wire systems, at least four parallel devices are required for military aircrafts and higher degrees of redundancy for civil aviation [Collinson, 1999].

[8]In a fault tolerant unit, a defined number of faults does not lead to a failure of the unit [Wilwert et al., 2005].

some cases, a fault tolerant unit can also be constructed based on only one ECU if the ECU features a multi-core architecture and an appropriate board design in combination with special mechanisms to allow the execution of multiple safety critical (and possibly also non-safety critical) functions independently on this platform. Appropriate strategies are, e.g., outlined in the RECOMP project [Motruk et al., 2012] funded by the Federal Ministry of Education and Research (BMBF) or by [Philipps, 2012]. Amongst others, a dedicated software functions and storage management has to avoid failures of multiple functions due to a common cause[9].

Network centric approaches

A further approach to reduce the number of redundant ECUs is a network centric architecture: in this case, the distributed nodes monitor each other. If a single ECU fails, the other ECUs react by adapting their mode of operation. As a result, the number of ECUs can be reduced, but the complexity of each individual device increases [Kelling and Heck, 2002; Johannessen et al., 2002; Sakurai et al., 2008]. The cost efficiency of such a solution has to be evaluated for each specific application. For example, Brake-by-Wire systems typically benefit from a network centric approach if each brake is set up as independent unit, which is capable to coordinate with the other brakes.

Sensor and actuator redundancy ❸

For the sensors in the vehicle, similar redundancy strategies as for the ECUs have to be applied to ensure safe operation. Mostly, measurements are acquired with three sensors at the same time to allow a majority voting among the measurements and thus to detect faults as, e.g., demonstrated by Bertacchini et al. [2005]. To reduce hardware costs, mechanisms for analytical redundancy can replace one or two sensors by software algorithms. In the course of this, "virtual sensor data" or diagnostic residuals are generated for an investigated signal by reusing the other available sensor data and knowledge about the system [Anwar and Niu, 2010; Gadda et al., 2007; He et al., 2010; Kim et al., 2010]. The fault tolerance strategies can also be implemented mechanically or electronically within the sensors or actuators [Dilger et al., 2004] by using model based diagnostic algorithms to identify faults and to derive appropriate actions already within the units [Isermann and Beck, 2011; Muenchhof et al., 2009]. As a result, the required number of physically separated redundant units can be reduced. In general, for actuating the safety critical systems in the vehicle, redundant units have to be implemented. For the steering, mostly two redundant actuators are used at one axle [Heise, 2012; Muenchhof et al., 2009; Wilwert et al., 2005; Zhen et al., 2005; Zuo et al., 2005]. For the braking system, each wheel usually features an individual actuator [Isermann et al., 2002; Papadopoulos et al., 2001; Reichel and Armbruster, 2011]. The feedback actuators at the user input devices are frequently classified as safety critical. But, the criticality is lower than the one of the actuators at the wheels. Thus, developers implement these actuators as single actuators [Anwar and Niu, 2010], redundantly [Wilwert et al., 2005], or provide mechanical back-up feedback [Pruckner et al., 2012; Zuo et al., 2005] depending on the safety concept.

[9] A common cause failure is a "failure of two or more elements of an item resulting from a single specific event or root cause" [ISO 26262-1, 2011, p. 3].

Network ❹

To connect the electronic components in the vehicle, a proper networking strategy has to be chosen. The used data bus systems have to feature at least single redundancy including physical separation in wiring [Wilwert et al., 2005]. Some researchers even propose more than two physically independent channels, such as Sinha [2011] and Sundar and Plunkett [2006]. Additionally, the overall network has to support a precise timing of messages to ensure that lost or delayed messages are detected, and a maximal round trip time is guaranteed [Heiner and Thurner, 1998]. Wilwert et al. [2005] derive a maximal acceptable end-to-end response time for driver inputs to the steering actuators of 17.6ms[10]. Beyond this limit, the vehicle becomes unstable. In applications, exclusively time-triggered data bus systems are relied on, e.g., TTCAN [He et al., 2010], TTP/C [Papadopoulos et al., 2001; Blanc et al., 2009], or FlexRay [Sinha, 2011; Sundar and Plunkett, 2006; Waraus, 2009]. Some research projects also investigate the applicability of Ethernet in combination with a time triggered extension [Müller et al., 2011]. These bus systems provide precise and deterministic communication timings at the price of less flexibility for spontaneous adaptations during the design process [Mishra and Naik, 2005].

Operating systems ❺

Utilizing the data-bus timings, the safety-related applications within the network have to be synchronized to ensure defined latencies [Sundar and Plunkett, 2006]. Several operating systems support this task, such as a modified OSEK [Sakurai et al., 2008], OSEK Time, FTCom [Wilwert et al., 2005], or upcoming also AUTOSAR [AUTOSAR, 2012; Mitzlaff et al., 2010; Tucci-Piergiovanni et al., 2011]. Nevertheless, the applicability of each operating system has to be confirmed individually depending on the required precision of timings and the computational resources of the network nodes.

Functional redundancy ❻

Assuming proper interaction and operation of the individual components within the vehicle, functional redundancies between different actuators and in particular between different types of actuators (steering, drive units, brakes) can be exploited to achieve the safety goals. Few research projects are available that use these redundancies for a proof of functional safety in accordance with ISO 26262, but several projects investigate the available functional redundancies and cross-couplings between the individual actuators from a view of vehicle dynamics based on simulation or experimental vehicles [Arbitmann et al., 2011; Dominguez-Garcia et al., 2004; Euchler et al., 2010; Hayama et al., 2008; Johannessen et al., 2002; Reinold et al., 2010]. Further contributions from research groups in the field of vehicle dynamics as Gerdes's group at the Stanford University and Trächtler's group at the Universität Paderborn exploit the capabilities of highly flexible experimental vehicles to improve vehicle handling during critical driving situations while assuming a proper operation of the Drive-by-Wire system. These research results serve as a guideline for what vehicle control algorithms are already available or can be expected to be available in the near future.

Power supply ❼

As a basis for the safe operation of the Drive-by-Wire system, a fault tolerant power supply unit is mandatory. Typically, systems with redundancy and mutual

[10]For comparison: In aviation, the sensors are sampled about 100 times per second, which roughly equals the minimal demands in the automotive field (data for an A320, [Collinson, 1999]).

isolation are implemented [Abele, 2008; Kelling and Heck, 2002; Sieglin, 2009]. The GM vehicle Sequel implements double redundancy and an additional central unit to reconfigure the power supply in case of failure [Sundar and Plunkett, 2006]. Such reconfiguration units for power supply systems are frequently implemented [Armbruster et al., 2006; Sundar and Plunkett, 2006], but may also turn out to be a weak spot of the architecture due to their huge impact on the overall system.

To monitor the overall system, a suitable diagnostic unit or function has to be implemented. These units have to ensure that occurring faults are detected such that the remaining system can be reconfigured to maintain sufficiently safe operation. According to the state-of-the-art and typical legislative requirements (ECE R13), the system has to tolerate at least one independent fault and still maintain (degraded) operation [Armbruster et al., 2006; Pruckner et al., 2012]. Most components of a Drive-by-Wire system already provide local diagnostic functions and output the results of these functions. Additionally, information can be extracted by network overarching monitoring mechanisms for timings and interfaces. To derive suitable actions from this information, different approaches, mostly relying on heuristics and probabilistic mechanisms, are typically applied [Bergmiller et al., 2011b; Isermann and Beck, 2011; Muenchhof et al., 2009; Schwall, 2005; Schwall and Gerdes, 2002]. The challenges for these algorithms are to guarantee short execution times and to provide traceable decisions, which renders most machine learning based approaches unsuitable. Typically, the approaches regard the vehicle as not self-healing. Thus, repair of defective components is not performed online. In aviation, a restart of components is considered to "heal" the system and improve the functional safety [Schroer, 2008]. This idea is also investigated theoretically for the automotive domain [Pimentel, 2003] but barely followed for the safety critical systems in real vehicles.

Online diagnostics, degradation ❽

Finally, the individual pieces of information on the system acquired by the diagnostic system can be related to each other and collected in a "knowledge base", which serves as a basis for long-term decision making while driving. As a result, critical situations can be avoided or handled more efficiently. Sec. 8.2 will provide a detailed discussion of the associated state-of-the-art and outline the approach chosen for MOBILE.

Knowledge base ❾

To conclude the overview of the state-of-the-art, a brief differentiation of MOBILE and a summary of the lessons learned from the state-of-the-art are provided. Also, two closely related research projects are highlighted. As was shown, multiple research groups are actively working on Drive-by-Wire systems, but few groups investigate the overall system including steering, braking, and the propulsion system with regard to flexibility, mutual dependencies, and functional safety in a real vehicle. Usually, only single systems are investigated on a test-rack or in simulation. Some research groups as the ones of Gerdes and Trächtler [Beal and Gerdes, 2010; Gadda et al., 2007; Trächtler and Niewels, 2006; Reinold et al., 2010] have built vehicles with extended by-Wire functionality but currently focus on vehicle dynamics rather than the functional safety of the EE system. Research results from these groups assist to identify the functional redundancies in MOBILE.

Differentiation of MOBILE

SIRIUS 2001

The approach followed by Johannessen [2001] for the "SIRIUS 2001" (Sec. 3.1) is the one that most closely resembles the approach in this thesis. The EE architecture follows a network centric approach and to some extent the functional redundancies are considered, e.g., steering by differential braking. The power supply is not considered for the functional safety evaluation. In a follow up project (FAR project), a model vehicle with four wheel steering and actuators to individually drive and brake the four wheels was built [Johannessen et al., 2004b]. The successful application of the network centric approach suggest exploitation of such an approach for MOBILE. Still, flexibility requirements for MOBILE and the need to consider the overall system including the power supply and a propulsion system capable of torque vectoring require reconsideration. In particular, the pure network centric approach in the SIRIUS 2001 can be expected to limit applicability as a tool.

SPARC project

The European project *SPARC* [Armbruster, 2009; Reichel and Armbruster, 2011; Sieglin, 2009] stands out by thoroughly investigating a full by-Wire concept for application in different vehicle classes (trucks and cars). The project presents a Drive-by-Wire architecture that is applied to different experimental vehicles with one steerable axle and a Brake-by-Wire system. The by-Wire system mainly sticks to single redundancy for the hardware components and includes a degradation approach to handle any faults. Still, the system architecture requires the memory in the network to be available in quadruple redundancy, as all nodes have to be able to perform all computational tasks. Compared to this project, MOBILE features a higher flexibility in actuation and targets an even lower degree of redundant actuators and controllers by exploiting the available functional redundancies instead.

Lessons learned

In general, MOBILE has to be designed as a functionally safe experimental platform and thus is subject to a unique combination of requirements in terms of flexibility and functional safety. Nevertheless, the design of MOBILE exploits several core principles outlined in the state-of-the-art:

- Any actuators or ECUs should be available at most in single redundancy, less devices are preferred if functional redundancies can be exploited.
- Initially, sensors are implemented in a double diverse redundancy for majority voting, as the analytical redundancy can easily replace existing sensors later.
- Time triggered communication facilitates precise synchronization of the applications throughout the network.
- If possible, safety critical components are kept independent from each other such that the vehicle can tolerate one independent fault and guarantee an emergency operation interval.
- Finally, force feedback is not regarded as a highly safety critical function, as a trained driver is driving MOBILE. As a result, the associated actuators are not implemented redundantly, but a mechanical open-loop feedback is provided.

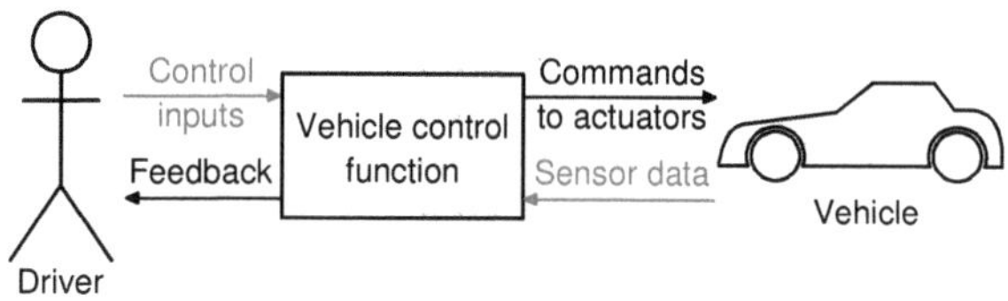

Figure 5.4.: Functional architecture of MOBILE at the "vehicle layer" in the style of a UML context diagram (coloring: orange = inputs, black = outputs)

5.2. Hierarchical Architecture Derivation

Taking the lessons learned from the state-of-the-art as a basis, this section resumes the architecture definition process at step 3 (Fig. 5.1). The architecture is intended as a template for by-Wire vehicles with the needs for flexibility and functional safety at low costs. These demands are addressed by relying on cutting edge research in the field of vehicle dynamics. As will be outlined at the end of this section, the basic design approaches reflected by the architecture can be transferred for cost efficient implementation of functional safety in series vehicles. As mentioned, the architecture will be derived top-down along the hierarchical layers introduced in Fig. 5.2. At each layer, at first the functional architecture is presented, then a suitable hardware architecture is derived that allows to execute all needed functions and fulfills the requirements for modularity and functional safety in the research context (steps 3-5 in Fig. 5.1).

5.2.1. Vehicle Layer

The fundamental functional architecture at the "vehicle layer" is dominated by the vehicle control function if other add-on functions are neglected (Fig. 5.4). The function acquires data from the driver (control inputs) and accordingly controls the actuators of the vehicle (commands to actuators). Vice versa, sensor data is gathered to execute the vehicle control function and to provide feedback to the driver. The hardware architecture is structured similarly to the functional set-up. It consists of the part of the EE system concerned with vehicle control. Individual hardware components are not distinguished at this layer. The functional safety requirements at the "vehicle layer" are associated to the vehicle control function.

5.2.2. System Layer

The "system layer" splits the vehicle control function into the most important components. Figure 5.5 outlines the resulting functional view. The design covers the special purpose of the vehicle as a development tool and the intention to exploit the functional redundancies:

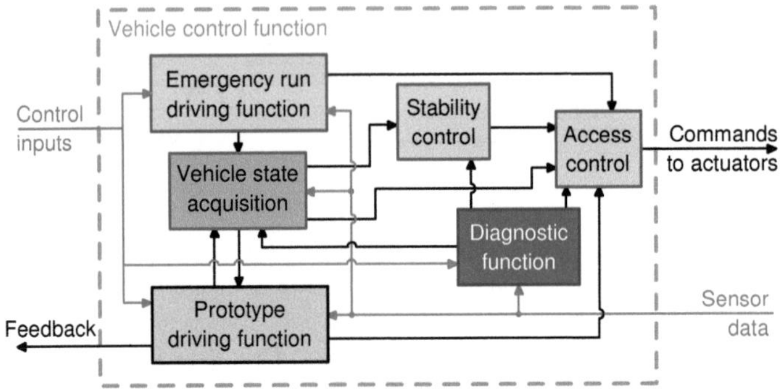

Figure 5.5.: Functional architecture of MOBILE at the "system layer" (colors for cross reference with following figures)

Functional architecture

A *prototype driving function* that is realized by software and possibly also hardware under development controls the vehicle actuators in experimental mode. When implementing this function, the developer is provided with full access to the vehicle including all sensor data and the actuators [→ R1, R2]. The functionality implemented by the prototype driving function can hugely vary, e.g., four-wheel steering vs. steering of only one axle.

As a basis for the prototype driving function and other control applications in the vehicle, the *vehicle state acquisition* function gathers all available sensor data related to the vehicle dynamics throughout the vehicle network and derives the current vehicle state. Thus, the state acquisition contributes significantly to the tooling character of MOBILE [→ R1].

Jointly, the prototype driving function and the vehicle state acquisition function enable the control of the vehicle. Further components are added to ensure safe driving. This includes an emergency run driving function, a stability control module, and the access control (colored red in Fig. 5.5). The *emergency run driving function* provides the basic "fall-back" control of actuators in a hot-standby manner. In fall-back mode, the actuators are operated in their most basic mode of operation relying on a minimal set of extra sensors. No cross-couplings or dependencies with other functions exist. Still, the emergency run driving function provides the driver full access to the steering actuators, brakes, and drive motors [→ R3].

As MOBILE is only equipped with one actuator for steering, braking, and drive at each wheel, functional redundancies have to be exploited for functional safety. Accordingly, the *stability control system* operates on the one side as a conventional stability control known from series vehicles. On the other side, it compensates the failure of a single actuator by adapting its control strategy [→ R4]. To distinguish failure states, the stability control includes data provided by the diagnostic system. It has to be pointed out that the stability control system, as will be detailed later,

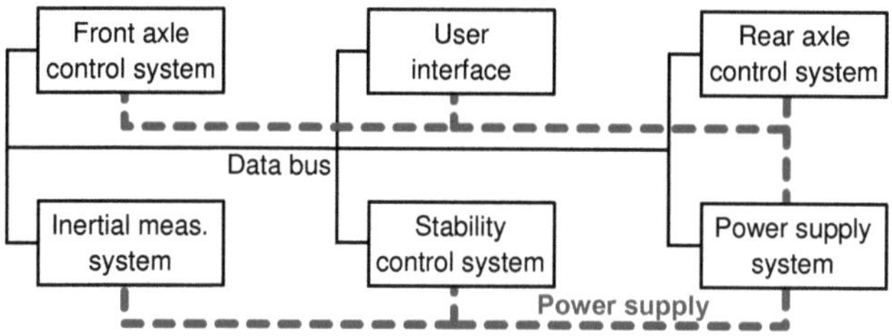

Figure 5.6.: Hardware architecture of MOBILE at the "system layer"

is not yet implemented on MOBILE, but is subject to ongoing research.

The *diagnostic system* monitors the current state of the vehicle from an electric/electronic point of view. It derives the failure probabilities of components and distributes the results throughout the network.

Finally, the *access control* determines which driving function may control the actuators: the prototype driving function, the emergency run driving function, or the stability control if an actuator fails →R3. The stability control re-uses the low-level basic actuator access to control the vehicle, which is implemented both by the emergency run driving function and the prototype driving function.

Hardware structure

The hardware units have to provide the basis for the execution of the given functions. Additionally, the independence of components reduces the number of common cause failures and increases the modularity of the vehicle. Figure 5.6 outlines the resulting architecture at the "system layer". The vehicle state acquisition function and the stability control function are implemented on the individual hardware units (*inertial measurement system* and *stability control system*) with no redundancy. A failure of one of these units cannot induce a total system failure if fail-silent behavior is ensured, because the emergency run driving function remains unaffected. The driving functions (emergency and prototype) are distributed over three control systems. The *front* and *rear axle control systems* control the actuators at the front and rear axle and pre-process sensor data acquired at each axle. The user inputs are acquired by the *user interface* and provided to other units via a data-bus. This contributes to fulfill the modularity requirements →R1. For example, a modification to an axle only requires adaptation of the associated control system. An additional *onboard computer* can be added to the hardware set-up for complex calculations required by the prototype functions. As this computer can always safely be disconnected from the network, it is not further considered in the hardware architecture. A *power supply system* provides the required electrical energy to all mentioned systems and can also cut power in case of failures. This contributes to the fail-silent behavior of the supplied components.

Merging views

Figure 5.7 shows the distribution of the diagnostic and safety related functions across the network. This design supports the modularity and reduces the probability of failure of the overall control function. For example, individual modules can

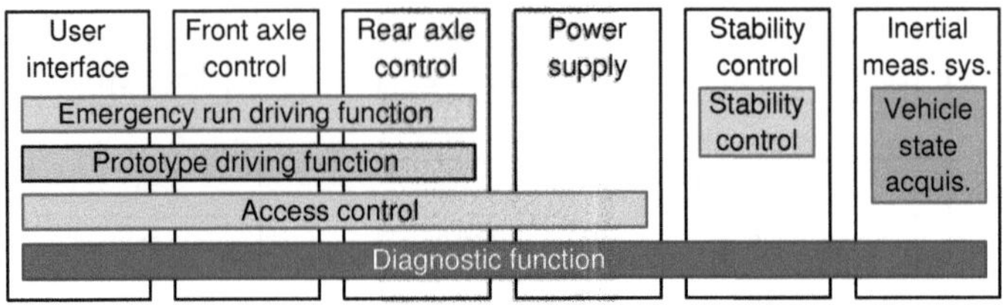

Figure 5.7.: Associating functional elements to hardware units at the "system layer"

easily be exchanged, as low-level tasks are kept local, while the system is, by its design, able to maintain at least a degraded operation if one unit fails.

5.2.3. Subsystem layer

Driving functions

Further detailing of the functional architecture down to the "subsystem layer" reveals the important functional modules and their interaction (Fig. 5.8). For the driving functions, the *data acquisition*, the *data processing*, and the *actuator control* are distinguished. If necessary, the access control function blocks actuator access rather than starting or stopping the execution of a function. Thus, the data acquisition and processing is performed in a hot-standby manner ensuring short switching times.

State estimation

The vehicle state acquisition is split into the *inertial measurement* with an appropriate sensor platform, the *sensor data fusion*, and the *state estimation.* The sensor data fusion and the state estimation combine the inertial measurements with classical sensor data, such as wheel speeds or steering angles, and estimate unmeasurable values, such as the side slip angle (Sec. 4.2.3). Depending on the mode of operation, data is acquired from the emergency run function or the prototype driving function.

Stability control

The stability control has to generate a *reference behavior* for the vehicle based on internal models of the desired vehicle dynamics. If the vehicle deviates from the reference, the stability control has to intervene. Depending on the current task of the stability control, different reference models are referred to. For classical stability control, a model approximates the stable handling of the vehicle. For later safety evaluation, e.g., also a simple bicycle model with limited dynamics can define the safe state. The control system then has to be powerful enough to guarantee this minimal performance even if an actuator fails.

Diagnostic system

The diagnostic system extracts relevant information from the *hardware monitoring* algorithms and the *globally available data* in the vehicle. Hardware monitoring relies on classical diagnostics as included in the state of the art (Sec. 5.1). The global data acquisition takes into account the driver commands and the reaction of the vehicle to these commands from an EE view. If deviations are detected, the diagnostic system derives which units might be faulty. More details on the diagnostic algorithms are given in Sec. 7.1.

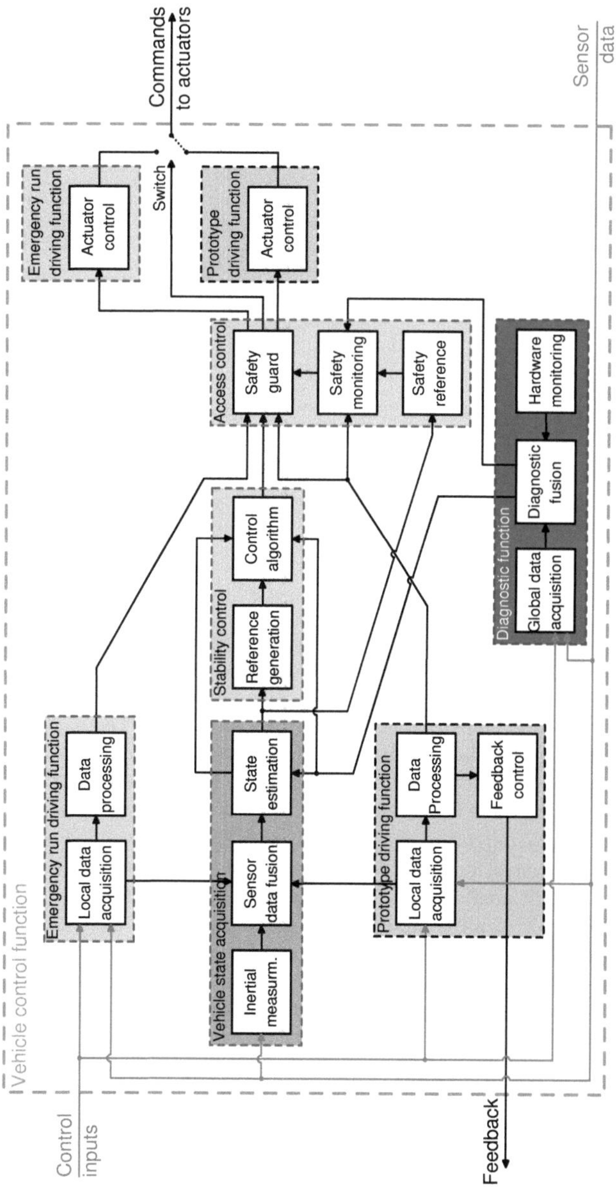

Figure 5.8.: The functional architecture of MOBILE at the "subsystem layer"

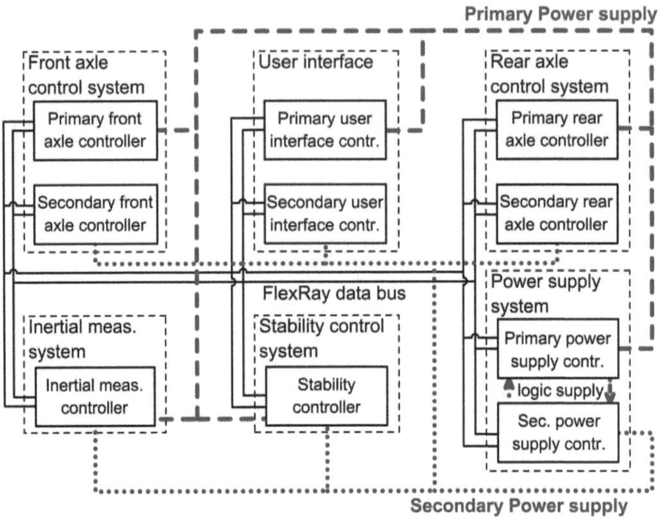

Figure 5.9.: Hardware architecture of MOBILE at the "subsystem layer"

Access control

Access control is split into the *safety reference* generation, the *safety monitoring*, and the *safety guard*. These functions re-configure the system if a driving function fails. Basically, the operation is similar to the one of a "stability control" for the EE system: a safety reference describes the desired state of the EE system, and the safety monitoring identifies failures and triggers the appropriate actions, which are then executed by the safety guard. On MOBILE , these functions are implemented based on state diagrams fed by the inputs from the diagnostic system.

Power supply

The hardware architecture at the "subsystem layer" is outlined in Fig. 5.9. To avoid a loss of control due to a loss of power, the power supply system features redundancy. Additionally, the *power supply controllers* contribute to the desired fail-silent behavior of all components in the vehicle. If a controller is classified as defective, the power for the controller can be cut. Thus, a fail-silent behavior can be enforced externally if internal mechanisms fail. Within the power supply system, the two power supply controllers supply the logic part of each other. As a result, a malfunctioning power supply controller can be switched off by the partner controller. The safe state of the controllers is to supply all connected controllers. Using this configuration, all reasonable failure scenarios of the power supply can be handled in cooperation with decentralized safety guards executed on each node.

Stability controller

If only the failure of one unit has to be tolerated, the stability control and the inertial measurement unit need not be fault tolerant and thus are implemented on only one controller each (the *inertial measurement controller* and the *stability controller*). Still, each unit is powered by both power supply lanes. This is necessary, because a total loss of one power line would lead to a loss of all connected controllers, two diagonal brakes, the propulsion at one axle, and two diagonal steering motors. As a result, the stability control system is needed to maintain a minimal degree of controllability. If the stability control and the state acquisition are powered exclusively by the faulty lane, failure compensation would not be possible. If the stability control itself fails, power can be cut by the power supply systems to ensure fail-silent behavior.

Fault tolerant units

A failure of the units executing the driving function would lead to a full or partial loss of control. Thus, these systems are set up as fault tolerant units consisting of two controllers, which are each powered by a different power supply. Within each fault tolerant unit, three diverse sensors for acquisition of the user inputs and the actuator positions are available, while actuators are only available once for each task. The wiring of the sensors and actuators to the axle controllers is driven by the requirements of the components available on the market. Mostly, CAN-bus connections implementing a CANopen protocol are relied on. Furthermore, some digital and analog sensor signals are evaluated and directly connected to the axle controllers. Within each axle, the allocation of the sensor signals and the actuator commands to bus systems ensures that the stability control unit can continue to control the axle such that at least neutral behavior with regard to vehicle dynamics can be achieved if a single bus fails. E. g., the drive motor of one wheel is connected to a different bus than the braking unit mounted on this wheel. Thus, if one system fails, the wheel can still be decelerated to some extent – either by recuperative or by mechanical braking. Communication between the two controllers within a fault tolerant unit and among the fault tolerant units is performed via a time triggered double-channel FlexRay data bus.

Merging views

Merging the hardware and functional view reveals the allocation of functions to hardware components (Fig. 5.10). It becomes obvious that the front and rear axle modules are set up analogously: the primary controller associated to an axle executes the prototype software under development and controls all actuators including the feedback devices. The primary controllers only implement a simple safety guard that can block the booting of the node if commanded by the power control units. Additionally, the basic diagnostics for the underlying hardware are available. These diagnostic algorithms are not visible to the user and are executed in the background. This enables the developer to act almost unrestricted by the safety concept.

The main diagnostic functions are implemented on the secondary controller. The secondary controllers are operated in hot-standby manner to quickly take over the vehicle control if required. While not controlling the vehicle, the available resources are used to execute appropriate diagnostic algorithms that will be discussed in more detail in Sec. 7.1. These algorithms continuously compare the actions of the primary controller to a safe reference and consider the results of the hardware monitoring units on the primary and secondary controller. If one node fails, the secondary controller communicates the failure state and a proposed action to the power supply controllers either by a dedicated message or by falling silent.

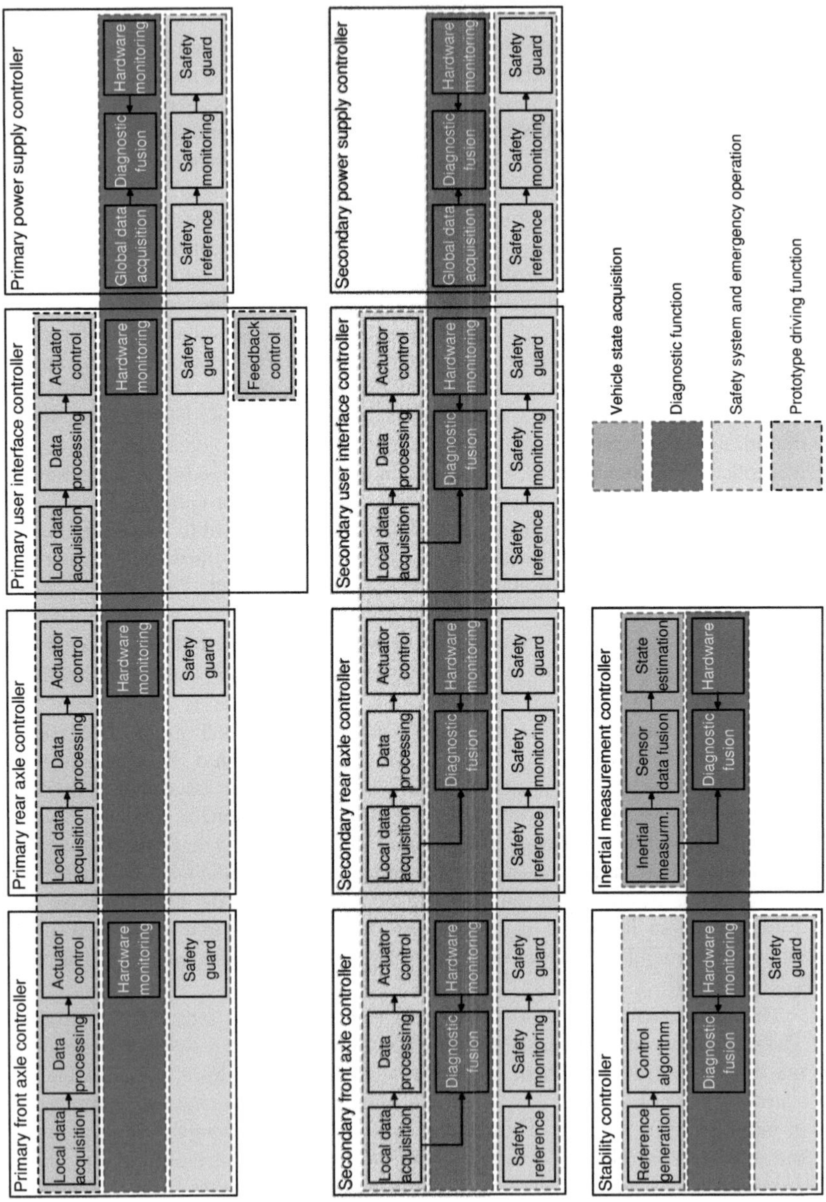

Figure 5.10.: Merging hardware and functional aspects at the "subsystem layer"

The power supply confirms the requested action if appropriate or cuts power to a defective node if required. It is important to note that the secondary controller continuously has to identify the basic mode of operation of the primary controller in order to achieve a smooth taking over after a failure. For example, the steering ratio is adapted online. Still, the adaptation is performed within strict bounds to avoid adaptation to an erroneous behavior. If the secondary node is subject to failure, a restart or reinitialization can be triggered to repair the system. For the primary node, such measures are not applicable, because the prototype software might go through unintended initialization routines while driving. Obviously, the secondary controller represents a possible source of critical single point faults, because the node is responsible for the decision making within an axle. This challenge is addressed by a network centric monitoring approach. The distributed diagnostic system monitors proper operation of all nodes by an Alive Network Management Vector based mechanism similar to the one implemented in the bus controllers of TTP/C [Sakurai et al., 2008]. For MOBILE, the alive monitoring is implemented at the application layer.

As indicated, the power supply controllers basically implement the access control function and thus play a vital role for coordinating the safety activities. To prevent scenarios where a single power supply controller can cause the system to fail, the power supply nodes monitor each other intensely and include information from the network wide alive monitoring. Additionally, the individual nodes in the axle modules perform consistency checks between the commands of the two power supply units.

With the presented view of the "subsystem layer", the hierarchical introduction of the hardware and functional aspects concludes. The following hierarchical layers ("component" and "elementary layer") feature further increasing levels of detail and focus on individual functions and their allocation to hardware parts within one controller similar to the approach presented by Abele [2012]. To some extent, these investigations have been made in the MOBILE project during the layouting of the electronics and implementing the software. Still, several hardware parts are sourced externally and no detailed information is available.

5.2.4. Software View

To conclude the architecture introduction, a brief look at a software view of the system is taken (Step 6 in Fig. 5.1), which covers aspects that have not been touched by the functional views introduced so far. Basically, each introduced function has to be implemented in some way as a software function. Still, the overall software structure "orthogonal" to the demonstrated application layer was not yet considered. This structure includes the layered approach from hardware abstraction to the application modules and the organization of their interaction. Figure 5.11 provides a simplified overview of the structure on each node of MOBILE. Every node runs a custom written "mini operating system" that fulfills the requirements for a basic task scheduling, low resource consumption, and little and defined latencies

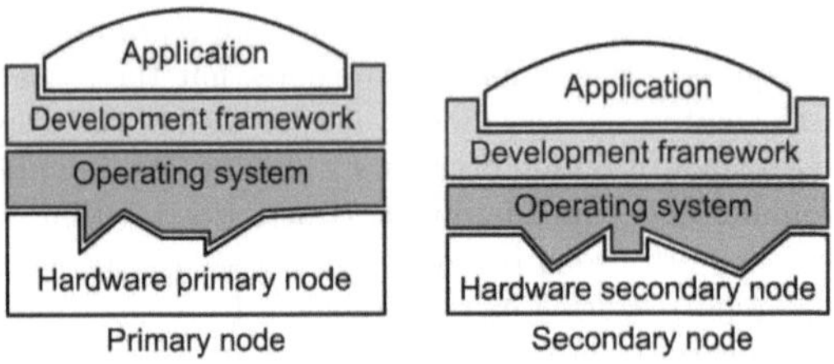

Figure 5.11.: The software framework on the network nodes

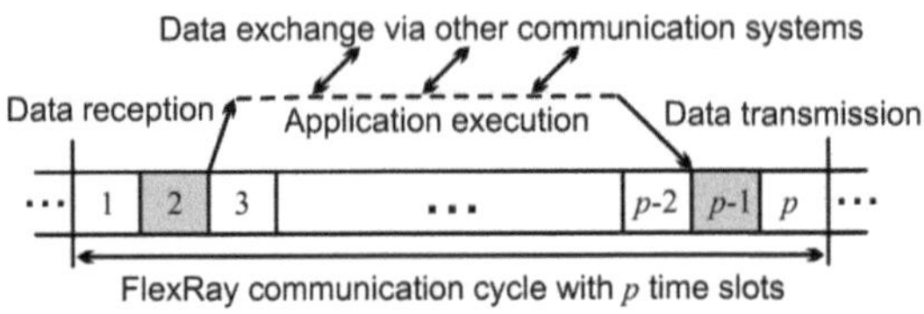

Figure 5.12.: The structure of the communication cycle within the network

while providing a full access to all hardware components. Also, it ensures that each node synchronizes itself to the global time basis provided by FlexRay. As a result, all actions within the individual nodes can be synchronized at a precision of microseconds if necessary. Then, in-cycle responses and cycle times down to a millisecond or below are possible. Based on the information given in Sec. 5.1 and the experiences gathered in the MOBILE project, this timing suffices for highly dynamic control of the vehicle. Figure 5.12 visualizes an in-cycle response scenario and the triggering of data processing. To reduce the probability of common cause failures, different operating systems are available for the primary and secondary nodes within a fault tolerant unit.

Application interface

The operating system interfaces with the code modules generated by Mathworks Embedded Coder (Sec. 4.1.3). This supports code reuse, and the graphical programming reduces coding errors. The application modules can be executed on any node within the network due to the common interfacing. In summary, the presented layering approach allows to flexibly exchange and reuse software modules, while at the same time hard real-time requirements can be met on microcontroller hardware → R2.

5.3. Summary and Critique of the Architecture

The proposed by-Wire architecture for an experimental vehicle was derived in a top-down manner to an extent possible within a research project. The architecture includes network centric and centralized approaches to bridge the gaps between

(a) the flexibility and functional safety by the network based monitoring combined with a suitable degradation concept and (b) between the functional safety and the costs by exploiting functional redundancies. Thus, the architecture can fulfill the initially set requirements for flexibility, functional safety, and the reduction of hardware redundancies (Tab. 4.1 in Sec. 4.1). Several key aspects distinguish the approach (step 7 in Fig. 5.1):

- The vehicle control function is implemented as a highly integrated system of all driving functions. This keeps the required hardware redundancies low (Sec. 5.1) and makes it possible to exploit the functional cross couplings between the different actuators for functional safety. Thus, the stability control system turns into an indispensable part of the architectural approach and facilitates economizing of the redundant actuators. As a result, all available actuators enable novel functions to create customer benefits while also contributing to the functional safety. Comparable systems with a focus on safety and a similar actuator set up are not found in the literature.
- The architecture strongly emphasizes the importance of the distributed execution of safety critical diagnostics at the application level to achieve low failure rates while keeping the degree of redundancy low. This conforms with the approach taken by Johannessen [2001], but additional centralized aspects are added to coordinate components and to increase the usability (power distribution units, onboard computers).
- The online reconfiguration and the "online repair" of the components as considered in aerospace, e.g., by reinitializing components, is supported by the architecture but not yet considered in the safety analysis presented later on.
- An efficient way to achieve fail-silent behavior of the individual network nodes is implemented based on the joint action of the power supply controllers and the decentralized safety guards.
- The architecture supports the flexible development of prototype software on the primary controllers and the onboard computers with little restrictions due to the safety mechanisms. Sophisticated monitoring algorithms help to keep the safety mechanisms out of the application software by performing "external monitoring". This approach might also be extended to complex functionalities in series vehicles. Then, not the complex function itself but the external monitoring system has to comply with the given safety requirements [Schäuffele and Zurawka, 2013, p. 212]. If done properly, this external safety guard can be structured generically and be re-used for different versions of the complex function.
- An online evaluated vehicle model clearly defines the safe state and serves as a baseline for the functional safety analysis. As a result, well-defined and meaningful requirements for proof of functional safety are defined at a vehicle level moving away from less meaningful requirements for the behavior of the individual components as they have been used so far in research and industry.

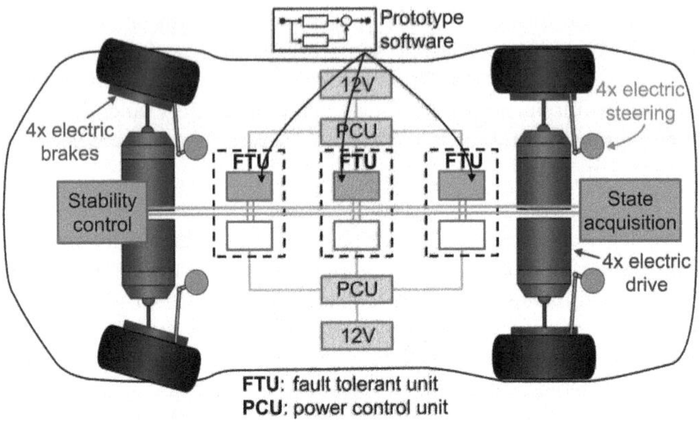

Figure 5.13.: Simplified architectural overview of MOBILE

Assumptions and future challenges

The introduced architecture is based on important assumptions that are still subject to research. To start with, appropriate fault detection and action derivation algorithms that consider the overall system state have to be derived. These algorithms implement the degradation concept of the vehicle and need to regard all given components including their own execution platform as possible sources of failure. A possible approach will be introduced in Sec. 7.1. A final parametrization of the algorithms for MOBILE that covers all relevant situations is part of ongoing research. Moreover, the stability control to exploit functional redundancies represents a key for economizing redundant actuators and thus the key to one of the main benefits of the presented architecture. Although quantitative data on the performance of novel stability control systems is still missing, results of multiple research projects in vehicle dynamics hold out that future stability control systems will be able to handle failures of individual actuators given over-actuated vehicles such as MOBILE (Sec 7.2). So far, MOBILE does not feature a stability control system as demanded by the architecture.

Summary

In summary, the architecture enables the construction of a powerful development platform that features both a good usability and the required flexibility while facilitating a sufficient level of functional safety. Novel applications and components for highly flexible vehicles can be evaluated during real test runs, and new mechanisms to ensure functional safety based on the functional redundancies and the vehicle stability control algorithms can be developed. Still, it must not be forgotten that the architecture is intended for a research vehicle and first versions of most of the individual components were developed in this thesis but are still subject to research. Also, conformance of the architecture with traditional structures and processes in industry was not considered by the functionally driven hierarchical architecture derivation. Figure 5.13 concludes the section by condensing important aspects of the vehicle architecture for quick reference in the following chapters.

6

A Tailored Approach to Functional Safety Evaluation[1]

"The ability to simplify means to eliminate the unnecessary so that the necessary may speak."

Hans Hofmann

"A person who, intentionally or negligently, unlawfully injures the life, body, health, freedom, property or another right of another person is liable to make compensation to the other party for the damage arising from this."
- German Civil Code[2] -

The above excerpt of the German Civil Code transferred to the automotive industry stresses the duty of any car manufacturer and engineer to ensure that their products are designed according to the state-of-the-art in terms of safety. If a vehicle demonstrably fails due to negligent design, the responsible person can be held liable for any consequences and thus has to provide compensation. As pointed out in Sec. 1, the proof of a state-of-the-art design of modern vehicles at a functional level is becoming more and more challenging. To give a guideline for the proper design and validation of the functional safety of new vehicles, ISO 26262 for functional safety in vehicles was derived from the more general IEC 61508[3] on the functional safety of electronics. As a result, the guidelines given in ISO 26262 would also serve as a benchmark for the design of the vehicle and the design process in case of law suits. Key aspects of ISO 26262 are the hazard analysis and the resulting classification of the derived safety goals[4] in terms of ASILs[5]. Amongst

[1] Parts of this chapter have been pre-published by the author in Bergmiller [2013].

[2] Official Translation by the Langenscheidt Translation Service of the German Civil Code (BGB) §823 in the version of its promulgation from 2nd of January 2002, last amended by the statute of 28th of September 2009.

[3] IEC 61508: Functional Safety of Electrical/Electronic/Programmable Electronic Safety-related Systems [IEC 61508, 2010]

[4] A safety goal is a "top-level safety requirement as a result of the hazard analysis and risk assessment" [ISO 26262-1, 2011, p. 14].

[5] Automotive Safety Integrity Levels, ranging from A (least stringent) to D (most stringent) [ISO 26262-1, 2011, p. 2]

P.J. Bergmiller, *Towards Functional Safety in Drive-by-Wire Vehicles*,
DOI 10.1007/978-3-319-17485-3_6

others, these levels determine the upper thresholds for the acceptable failure rate of the hardware underlying the examined function. In particular in the fields of electric vehicles and by-Wire approaches, ISO 26262 poses highest demands on newly developed systems. For these vehicles, classical ways to achieve functional safety by mechanical overdesign are no longer viable, as electronics control the vehicle.

Functional redundancy

Highly integrated systems, such as the one proposed in the previous chapter, and enabled by by-Wire control [Pruckner et al., 2012] promise enhanced customer benefits while limiting production costs. Nevertheless, the safety evaluation of such systems is time consuming and prone to errors [Papadopoulos et al., 2001] if the classical approaches outlined in ISO 26262 are followed. Also, the functional redundancies immanent to the system have so far barely been taken into account for safety evaluation. To address these challenges, a hierarchical approach to analyze the safety of the "driving functionality" is introduced in this chapter. The approach particularly considers functional cross compensations between safety critical systems that are traditionally examined separately, e.g., the braking, steering, or drive system. It will also be pointed out that to some extent such systems and approaches to achieve functional safety are not sufficiently addressed by ISO 26262. Due to its flexibility, MOBILE serves as a suitable platform to demonstrate the applicability of the hierarchical approach. The detailed evaluation results will be summarized in Cha. 9 after all the important features of MOBILE have been introduced.

6.1. Requirements based on ISO 26262

Starting from the item[6] definition being the overall vehicle control system in this case, a hazard analysis is carried out to "identify and to categorize the hazards that malfunctions in the item can trigger" [ISO 26262-1, 2011, p. 6]. These hazards are then associated to safety goals. Safety goals again serve to derive the technical and functional safety requirements for system components. Each safety requirement inherits the ASIL classification from the safety goal and thus from the identified hazards unless ASIL decomposition[7] is applied. If this is done for a highly integrated Drive-by-Wire system as introduced for MOBILE, one notes that all the requirements at the "vehicle level" that are not associated to the emergency-off system have to be associated to the overall vehicle control function. An association of requirements to clearly separated sub-functions may be possible in the functional architecture but is no longer useful if the hardware architecture is taken into account. For example, one hardware unit contributes to the braking, steering, and propulsion function. Figure 6.1 illustrates such a system structure at the "vehicle level" for an experimental vehicle like MOBILE with an emergency-off system and

[6]An item is a "system or array of systems to implement a function at the vehicle level" [ISO 26262-1, 2011, p. 10].

[7]According to ISO 26262 ASIL decomposition denotes the "apportioning of safety requirements redundantly to sufficiently independent elements" [ISO 26262-1, 2011, p. 2].

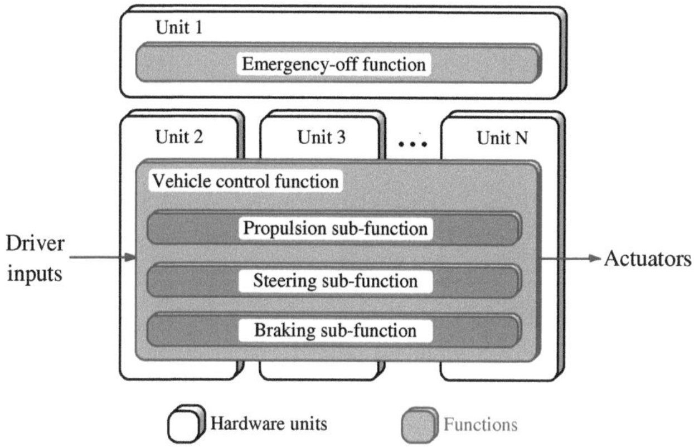

Figure 6.1.: Highly integrated system

a highly integrated vehicle control system based on Drive-by-Wire. The given system consists of several individual hardware units that are combined to fulfill the overall task. A similar tendency towards the integration of multiple safety critical functions on one control unit and within one mechatronic component can be observed in modern series vehicles, e.g., if control systems for longitudinal, lateral, and vertical dynamics are merged [Koehn et al., 2006; Smakman et al., 2008]. In research, Freitag and Kuhn [2012] go even further and propose to replace conventional brakes at the rear axle with an in-wheel motor that drives and brakes the wheels. Thus, borders between classically separated functions and systems start to blur. A safety evaluation highlights that the failure of one unit may lead to the loss or degradation of multiple functions.

Transferred to the simplified system structure given in Fig. 6.1, the sub-functions merge into the overall vehicle control function. As a result, all safety requirements have to be assigned to the one vehicle control function, and thus to the overall underlying hardware, which consists of several hardware units. According to ISO 26262, the highest safety requirement that is derived for one of the executed functions would then be assigned to the overall system. As a result, safety evaluation can lead to lower safety requirements for the overall system than intended if one unit executes multiple functions of lower safety criticality that directly or indirectly effect vehicle handling. This renders the previous hazard analysis based on the analysis of classically separated functions wrong, and thus it requires a re-evaluation based on knowledge about the evolving system architecture. The hierarchical approach outlined in the following assists such an evaluation of complex integrated systems by performing a structured top-down analysis.

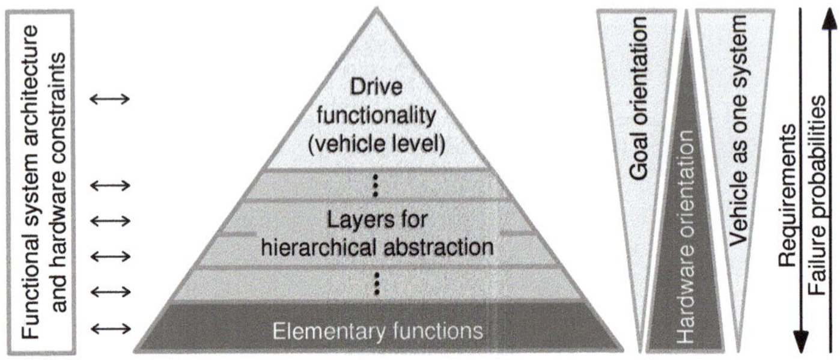

Figure 6.2.: The hierarchical approach to functional safety analysis

6.2. The Approach to Functional Safety Analysis

The proposed method for tailored hierarchical safety analysis presented in this section extends existing approaches in a number of aspects. The hierarchical approach particularly focuses on integrated systems with a *high degree of functional redundancy* and supports the usage of these redundancies for safety purposes. To reduce work effort for the analysis, the presented method introduces virtual systems and generalized failure states that support the developer to *focus on the analysis of the necessary components* by front loading knowledge about the dependencies in the system. The result of the analysis is a *failure rate for the overall system* that takes the required *emergency operation interval* for the vehicle into account. Knowledge about the available emergency operation interval is vital to ensure the safe stopping of the vehicle or to define a "limp-home" mode. To assess the performance of the distributed diagnostic algorithms in the vehicle, the approach also provides the diagnostic coverage of a globally operating *virtual diagnostic unit*. This diagnostic coverage can guide the further development of the local monitoring algorithms and encourage "vehicle level thinking" for diagnostics.

Perception of the system

Figure 6.2 illustrates the perception of the vehicle adopted by the hierarchical approach and includes important principles. The analysis is carried out at different hierarchical levels. In general, quantitative results and failure probabilities are propagated bottom-up, while dependencies of components and requirements are forwarded top-down. The functional and hardware architecture of the examined vehicle are considered for all hierarchical layers. The hardware influence diminishes as the hierarchical level increases, but the required understanding of the overall vehicle by the person in charge grows. Both profound knowledge about the interconnection of vehicle components and vehicle dynamics are needed at the higher hierarchical levels. As an example, Fig. 6.3 shows the hierarchical layers that were defined for MOBILE. As can be seen, the levels match the levels introduced previously for the architecture description. For the analysis of MOBILE, the

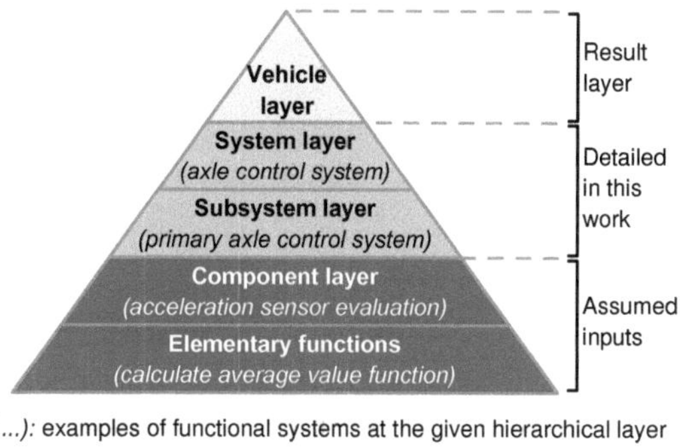

Figure 6.3.: Hierarchical layers defined for the safety analysis of MOBILE

"vehicle layer" is set to be the "result layer" as the overall vehicle control function shall be classified as being 'ok' or 'not ok', and the resulting failure rates shall be determined. The component and elementary layers are not covered in this thesis. Failure rates of these layers are assumed as inputs.

Starting from this perception of the system, Fig. 6.4 outlines the steps taken by the approach. Basically, the process starts with a targeted analysis of the vehicle architecture, then investigates relevant components, and finally performs the evaluation of the functional safety of the overall system. The following sections detail the individual steps using the example of MOBILE and review the related work.

6.2.1. Step 1: Define Hierarchical Layers

To start the analysis, the hierarchical layers are defined to support the mastering of complexity and to serve as a basis for the stepwise safety evaluation. The number of levels varies depending on the examined system. All layers up to the "vehicle layer" have to be detailed. Fig. 6.3 summarizes the introduced layers for MOBILE.

Hierachical layering in research

Such hierarchical layering of systems is frequently applied in research and industry to handle the complexity of automotive systems. E. g., Abele [2012] defines "vehicle level", "system", and "subsystem level" for the hierarchical derivation of safety requirements for the sub-functions and the components of a single ECU in an hybrid electric vehicle. Similarly, Papadopoulos et al. [2001] identify the need for a hierarchically structured approach to safety analysis using a Brake-by-Wire system as an example. Also for the analysis of human-machine interaction, hierarchical approaches are applied [Kirwan and Ainsworth, 1992].

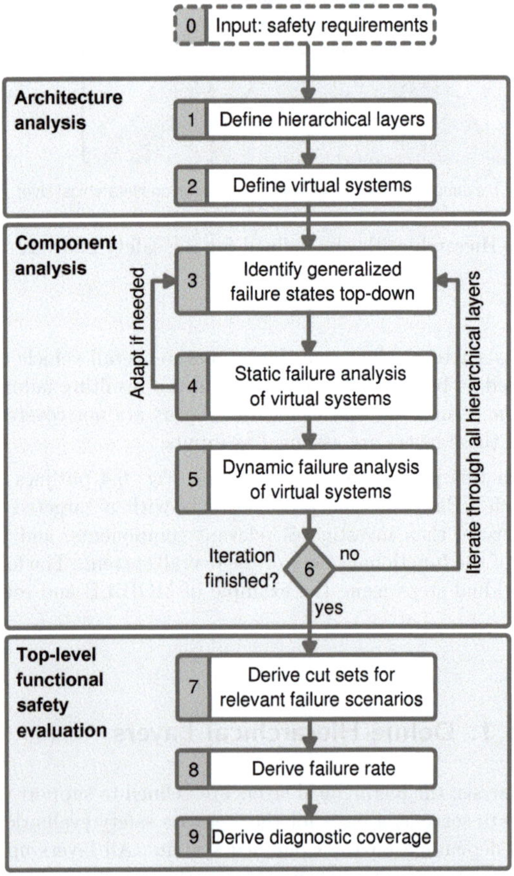

Figure 6.4.: Important steps of the hierarchical approach to safety analysis; the steps define the core process of this method

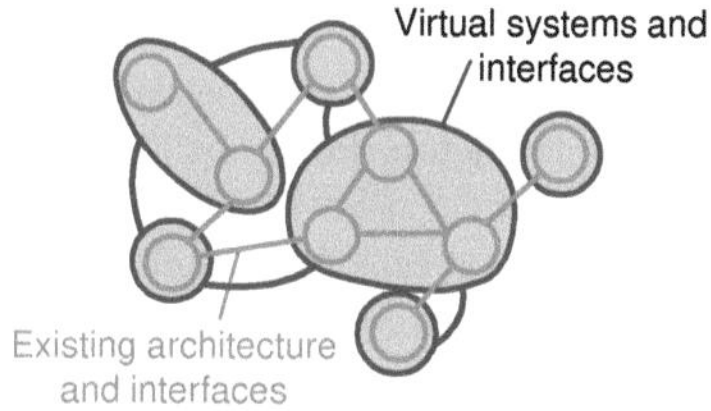

Figure 6.5.: Virtual systems introduced at a hierarchical layer

6.2.2. Step 2: Define Virtual Systems

To focus the analysis on safety aspects, a novel approach based on virtual systems is introduced. For the safety analysis, it is vital to determine which subsystems are subject to common cause failures, and which ones can be assumed to be independent. Splitting the overall system reasonably into these subsystems with regard to the safety goals reduces the complexity and the work effort. As a result, within each hierarchical layer, independent virtual systems with clearly documented interfaces are defined (Fig. 6.5).

Examples

To illustrate the meaning of a virtual system, some examples for MOBILE are given. In MOBILE, the front axle control system would be an example for a virtual system. This system includes all hardware and software units relevant to its function, e.g., the two microcontrollers and the software control functions running on these. In standard vehicles, typically other system definitions such as the braking system or the steering system are chosen, which could also have been done for MOBILE. Still, the further splitting of these systems would then quickly have revealed that choosing the systems like that causes huge effort for failure analysis as independent subsystems are harder to find. Nevertheless, this example shows that the system splits can be defined in the one or the other way and several ways may lead to a successful analysis of the system. That is why these systems are referred to as "virtual". The boundaries of these systems are not necessarily physically existent. Another virtual system for MOBILE, this time on the subsystem layer, would be, e.g., the secondary axle controller within an axle control system. This subsystem again includes all additional sensors exclusively used by the secondary axle control system and the according software algorithms.

Critique

The definition of virtual systems is challenging, as the developer requires profound knowledge of the functional, software, and hardware architecture at the given hierarchical level. Still, all following analysis steps can then be carried out based on the architecture of virtual systems without having to consider other architectural views. The virtual systems ensure linkage between these views, and overall complexity is reduced by front loading this knowledge. Typically, the first definition of virtual systems will be carried out middle-out[8], as independence between different units can be determined easier at the level of control units. Starting from there, the investigations are refined top-down. In this process, the boundaries of the virtual systems have to remain consistent throughout all hierarchical layers. Thus, a system on a lower layer must only be contained in one system at the next highest hierarchical layer. If that is not possible for a system, the system has to be moved up one level or the final level of granularity is reached for this component.

6.2.3. Step 3: Identify Generalized Failure States Top-Down

In step 3, the failure modes of the virtual systems are examined. As a basis for the analysis, the hierarchical approach introduces generalized failure states for each virtual system defined at a hierarchical layer. These states abstract information on the current failure state of the virtual system and include only the information needed by other systems on the same and in particular on higher hierarchical layers. This significantly reduces the work effort for later quantitative safety evaluations in comparison to existing approaches, such as the HiP-HOPS approach proposed by Papadopoulos et al. [2001]. Figure 6.6 provides examples of generalized failure states for a simple system. For the hierarchical approach, these generalized states serve as a well-defined and well-documented interface between experts working on different subsystems and possibly at different hierarchical layers. Thus, the developer of a component can focus on a locally well-defined work package, as the "vehicle level" effects have already been taken care of.

Definition process

Nevertheless, the definition of the generalized failure states is a challenging task and requires cooperation among experts. It mainly follows two principles. On the one side, the requirements from a higher layer have to be propagated top-down. This ensures that each state provides sufficient information to an expert working at a higher layer to evaluate the system based on the pool of underlying generalized failure states. On the other side, the structure and function of the virtual system influence the definition of the generalized failure states. This requires a sufficient understanding of the purpose and the behavior of the system in case of a failure. As a result, the generalized failure states of a virtual system have to be defined in cooperation of the experts on the hierarchically neighboring layers. This is then done layer by layer in a top-down fashion. For example, in the MOBILE project the expert on "vehicle level" has to agree on the generalized failure states of the power supply system with the person in charge of the power supply system at "system level". To reduce the work effort in the following analysis, the number of generalized failure states should be kept as low as possible, and the states should clearly relate to safety aspects. For example, in the MOBILE project only three failure states were derived top-down for each axle control system at the "system layer": 'destabilizing', 'neutral' or 'ok'. Thereby, the 'destabilizing' state means that the axle actively disturbes the vehicle dynamics meaning that it requires intervention

[8]In this context, a middle-out approach starts to define systems at a medium level of abstraction and then propagates the results both down to lower and up to higher hierarchical layers.

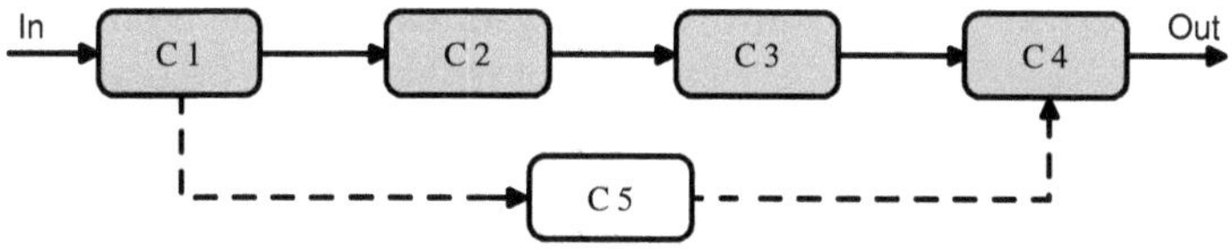

Failure of	Generalized failure state
C1 | C4 | (C5 & (C2 | C3))	Total loss of system
C2 | C3	Emergency operation
No failure | C5	Regular operation

Full operation: ⟶, Emergency operation: - -→,
Logic "or": |, Logic "and": &, Component x: Cx

Figure 6.6.: Simplified examples of generalized failure states for a system

by other systems not contained in this axle to make MOBILE follow the reference trajectory given by the internal simulation model (e.g., a wheel is locked). Analogously, the 'neutral' state indicates that the axle is not actively disturbing the dynamics but can be handled like a fixed rear axle (e.g., one brake unit fails, but the axle can still decelerate sufficiently using the other brake and the drive motors). In 'ok' state, the axle is fully operational. Of course, this definition of states may turn out to be not detailed enough when further investigations of vehicle dynamics control algorithms will be performed with MOBILE in future research, but they indicate the basic idea and serve as a basis for the example analysis in this thesis. The expert performing the failure analysis of the axle control system must then check when his system enters one of these states. If the expert cannot always make a clear association to a state, the generalized failure states have to be readjusted in cooperation between the two experts at the neighboring layers as pointed out before. E.g., a more detailed listing of failure states may be required for the impact of the axle control system on the dynamics of MOBILE. Such a re-definition can also be triggered after the detailed failure analysis in the next step (step 4) has been completed.

Related work

In the literature, the method of introducing generalized failure states has already been considered to some extent. Sinha [2011] defines generalized failure states for a braking system regarding one hierarchical layer. The states are not exploited for linking systems or to hierarchically propagate severity levels. An application of generalized failure states to a more complex system is outlined by Rehage et al. [2005]. The introduced states are identical for all systems ("active", "isolated", "active-hot", "passive-warm", "passive-cold") and applicable for aerospace systems with multiple similar systems in parallel redundancy. In the automotive MOBILE project, these fixed states are not suitable for considering functional redundancies.

6.2.4. Step 4: Static Failure Analysis of Virtual Systems

Taking the generalized failure states as a basis, a more detailed analysis of the virtual systems is required. Therefore, step 4 performs a "static" failure analysis without taking any timely effects into account. This analysis has to be performed on every hierarchical layer and targets the internal failure modes of each virtual system. The analysis of a virtual system is done based on the generalized failure states of its subsystems. Thus, detailing the analysis bottom-up is recommended. E.g. within an axle control system of MOBILE, the impact of failures of its subsystems as the brakes or the drive motors on its own generalized failure states ('destabilizing', 'neutral', 'ok') has to be analyzed.

Various methods support this analysis: ISO 26262-4:2011 suggests deductive, e.g., Fault Tree Analyses (FTA) as well as inductive analysis approaches, e.g., Failure Mode and Effect Analysis (FMEA). Details on FTA, FMEA, and further methods are given by Rausand and Hoyland [2009] or Löw et al. [2010]. This thesis mainly relies on a slightly modified FMEA that includes possible ways to diagnose and handle failures and a simplified FMEDA (Failure Modes, Effects and Diagnostic Coverage Analysis) to determine quantitative data. To derive quantitative data, the failure rates of software components (control and monitoring algorithms) are included. This extends the classical approach to failure analysis given in ISO 26262 that exclusively refers to hardware components for failure rates. In ISO 26262, the algorithms are only considered by the demanded diagnostic coverage to some extent. Software is considered to comply with ASIL requirements if the software development process followed the guidelines given in the standard. This approach may be valid for series vehicles with profoundly developed software components and little dependence on external influences. Still, the failure rate given for the vehicle according to ISO 26262 assumes perfect software and is only valid for the hardware set-up. For the experimental vehicle, the failure rate of software components has to be considered for two main reasons. Firstly, parts of the software running on the vehicle are prototypical and have failure rates that are several orders of magnitude higher than the ones of hardware components. These failure rates have to be estimated based on experiences in previous research projects. Secondly, algorithms for vehicle stability control are a vital part of the safety concept. These algorithms cannot be expected to operate properly under all environmental conditions or for all input configurations. The system, by its design, may just not be able to handle some rarely occurring situations. A failure rate has to be assigned that most likely has to be derived from statistical data acquired with a similar system already being used in vehicles. Another possible approach, similar to the diagnostic coverage analysis, could be to define failure rates for a catalog of typical software modules that are then combined for the overall software system.

The failure analysis, such as the FMEA, which is carried out during the hierarchical approach, benefits from the option to apply methods locally, which distinguishes the chosen hierarchical approach. Usually, FMEA has to include the global effects of a failure on vehicle handling, which makes the evaluation chal-

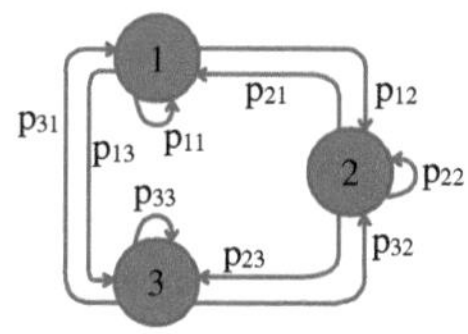

Figure 6.7.: A Markov Chain with three states

lenging for an expert for the local component. The hierarchical approach makes it possible to evaluate the failure effects with regard to the generalized failure states of a component, as the severity level needed for the FMEA is propagated top-down as mentioned in step 3. E.g., a loss of power in MOBILE is classified with highest severity level on the "vehicle layer". Thus, all generalized failure States within the systems and subsystems that lead to this failure are also assigned this level. As a result, the global context is taken care of.

6.2.5. Step 5: Dynamic Failure Analysis of Virtual Systems

Following the outlined static analysis, the probability of a virtual system being in each of its generalized failure states at a given point in time is derived. Therefore, an approach based on first order Markov Chains is taken. This procedure has already been suggested by Tkachev for the general "analysis of systems with complex structure" in 1983 [Tkachev, 1983] and was also followed by Zuo et al. [2005] to perform "quantitative reliability analysis [...] of Steer-by-Wire system[s]" in the automotive domain. ISO 26262-4:2011 references Markov modeling in general as a valid way to analyze the system design. A simple first order Markov Chain with three states is given in Fig. 6.7. The p_{ij} is the transition probability from state i into state j. According to Köhler [1983], this probability is defined as:

$$p_{ij} = P(X_{t+1} = s_j | X_t = s_i). \tag{6.1}$$

X_t defines the system state at the time step t and s_i is the feature vector characterizing the system state X within state i. As can be seen, the transition probabilities from one state into the other state at a given point in time only depend on the system state at the previous time step (Markov Property for first order Markov Chains). This is important for the failure analysis, as the history leading to a certain state does not explicitly have to be modeled but is contained in the structure of the Markov Chain. For the technical system, each p_{ij} is the failure rate of a part of a virtual system, and it is always derived bottom-up from the elementary hardware or software components on the lowest detailed layer. E.g., a simple bit flip in a register of one of the power control units of MOBILE can cause the power supply system on "system layer" to transition into the generalized failure state 'off'. The effect of such a low-level failure is traceable through all hierarchical layers due to the linkage via the generalized failure states. This is similar to the bottom-up propagation of failure probabilities presented by Papadopoulos et al. [2001]. In some cases, faults on the lowest layers may not become visible on the higher hierarchical layers, e.g., if the fault is treated at a lower level or is uncritical. Also, some state transitions at a higher layer can only be triggered by more than one fault on the lower layers. As all failure rates associated to hardware components vary over their lifetime due to aging, the Markov Chain is inhomogeneous and is therefore solved iteratively. The aging models can rely on statistical data from systems already in the market or models in the literature.

The hierarchical approach considers up to two independent faults, which may occur one after the other, to derive the Markov Chain for a component. This ensures that the reaction of the system to the first fault and the mode of operation afterwards can be evaluated in step 6. Further consecutive faults are not considered (see also the recommendation of ISO 26262-5 [2011], Annex C), which reduces the work effort. The failure rates and failure states of externally supplied components are directly fed into the system at the appropriate hierarchical layer.

After proceeding as outlined, a Markov chain with associated failure rates results for each virtual system at each hierarchical layer (Fig. 6.8 step a). It is important to note, that this dynamic analysis only has to be done on the lowest detailed layer for each component as the state transitions on higher layers result automatically from the linkage via the generalized failure states. Figure 6.8 only depicts one-way transitions, because the system is assumed to be *not self-healing*. Thus, a re-transition from one failure state into a state with less failures is not possible, unless it leads through the "ok" state "1", e.g., by a system restart.

The states of the depicted detailed Markov chain will typically not yet equal the generalized failure states of the system, but will correspond to the faults identified during the static failure analysis. Resulting, abstraction is needed, which is shown in steps b and c of Fig. 6.8. In this process, the states of the initially detailed Markov Chain are associated with the generalized failure states of the investigated virtual system. This should be possible due to the cooperative definition of the generalized failure states in step 3 and the static failure analysis with regard to the generalized failure states in step 4, which links the statically determined failure modes and the generalized failure states.

6.2.6. Step 6: Finish Iteration Through Hierarchical Layers

The evaluation outlined in steps 3, 4 and 5 is repeated throughout all hierarchical layers. As given in the individual steps, the virtual systems are defined primarily top-down, whereas the detailed failure analysis is then performed mainly bottom-up. Important additional information gained especially in step 4 that was not considered during steps 2 and 3 may require to, e.g., adapt some generalized failure states, which then triggers bottom-up re-evaluation of the system.

The last step in this iteration process is the analysis of the effect of each first and

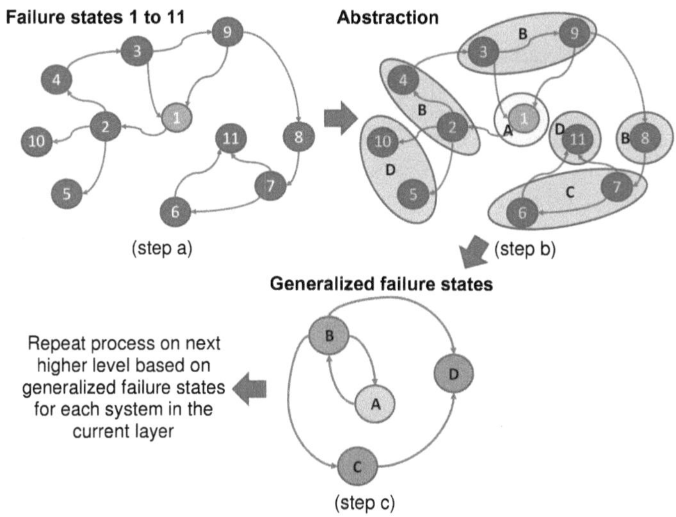

Figure 6.8.: A Markov Chain with states 1 to 11 and generalized failure states A to D resulting from abstraction

second failure on vehicle handling on "vehicle level" (step 4)[9]. E.g. for MOBILE, it has to be evaluated what effects the transition of one axle into the state 'neutral' has on the vehicle handling and what happens if additionally the stability control system fails. The following steps now describe the automatic derivation of the failure rates at the "system level" by numeric bottom-up propagation of failure rates. These failure rates then serve as a basis for the final calculation of the failure rate at the "vehicle layer".

For the simplified quantitative failure rate assessment in the MOBILE project, it is assumed that more than two independent faults are very unlikely to occur, and thus these scenarios are neglected, as it is commonly done in research [Armbruster et al., 2006; Johannessen et al., 2002; X-by-Wire Project, 1998] and by legislative prescriptions (ECE R13). The evaluation of a second fault is needed to determine, whether a sufficient emergency operation interval can be guaranteed after a first fault occurred.

[9]Step 5 will typically not be required on "vehicle level" as all quantitative data has already been gathered on lower layers and is propagated bottom-up.

6.2.7. Step 7: Derive Cut Sets For Failure Scenarios

After having completed step 6, all necessary inputs from experts are available to start the automatic quantitative safety evaluation. The evaluation algorithm is fed with the hierarchy of virtual systems, their failure states, and the associated transition probabilities. Then, the algorithm identifies which combination of failures of individual virtual systems causes the overall vehicle to fail. Therefore, two sub-steps are carried out in step 7:

1. *Virtually split the vehicle into one operational and one faulty part for each critical failure.* Based on the previous analysis, the critical failures of systems causing the vehicle control function of MOBILE to fail are identified by the algorithm. Each of these failures can be caused by a single-point/residual or dual-point fault[10]. After such a failure has occurred, from a "vehicle level perspective", the vehicle can be split into an operational and a faulty part. E.g., if one of the power supply units of MOBILE fails, this becomes the defective part. From a "vehicle level" perspective, it does not matter whether this is, e.g., due to a battery or an electronics failure. Thus, the algorithm now calculates the failure probability of this part. Therefore, it has to consider all possible single and double point faults that can occur within this part of the system and result in a failure at the "system level". This failure rate calculation is done in the next step. Of course, these calculations have to be repeated for all critical parts of the vehicle identified from the "vehicle level" analysis.

2. *Identify cut sets[11] for the non-operational part of the vehicle.* As mentioned, all relevant faults and fault combinations that cause a failure of a critical part of the vehicle have to be identified. Resembling the notion in reliability engineering, the algorithm identifies cut sets that cause a critical failure. For this, the algorithm considers combinations of up to two faults. Each cut set is determined on a hierarchical layer that provides a complete set of quantitative data to evaluate the effected virtual systems. Typically, this will be the lowest hierarchical layer. If the hierarchical approach is executed in parallel to the system analysis before all failure rates of low level components have been determined, the algorithm will indicate which failure rates have to be provided to determine the probabilities needed for a cut set. Thus, only data that is directly relevant for the safety analysis needs to be provided, which contributes to the cost reduction compared to approaches that start from a full set of quantitative data, such as the one by Papadopoulos et al. [2001].

[10] A single-point fault refers to a fault "in an element that is not covered by a safety mechanism and that leads directly to the violation of a safety goal"[ISO 26262-1, 2011, 16]. A dual-point fault leads to a failure in combination with another independent fault [ISO 26262-1, 2011, 6].

[11] "A cut set refers to the group of those elements or units which will make the system fail if their failure occurs. The minimum number of such units forms the minimal cut set" [Verma and Ajit, 2010, p. 85].

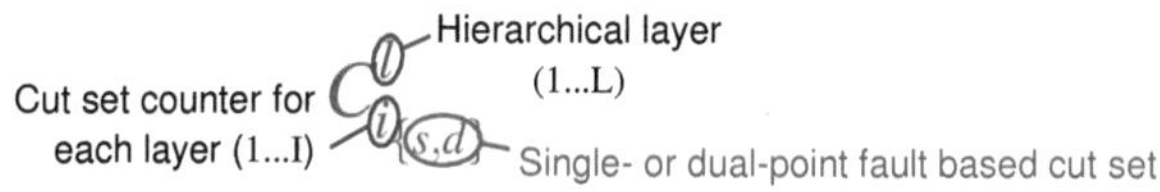

Figure 6.9.: Nomenclature for cut sets

Each identified cut set is referenced according to the notion introduced in Fig. 6.9. The superscript l, which indicates the hierarchical layer, and the cut set counter i uniquely identify a cut set. The indices s and d are supplementary to highlight whether the cut set is based on single-point/residual or dual-point faults. They constitute redundant information for better readability. If one of the indices or superindices is not given, a number of cut sets with all valid combinations for the omitted indices will be referenced. For example, C^l represents all cut sets at hierarchical layer l and C alone references all cut sets on all layers. For MOBILE, a cut set at "vehicle layer" would be the failure of the steering input at the user interface system. If this failure is generated due to a single-point/residual fault and the cut set is associated the 12th ID, the cut set will be referenced as $C^1_{12_s}$.

Example

Figure 6.10 illustrates a generic system with a system split derived from a critical failure as outlined in sub-step 1. Each virtual system is associated with the generalized states "o.k." or "not o.k.". The latter state is marked by the hatching. For MOBILE, the tree given in the figure would look similar, but more than two failure modes and more components would have to be depicted. In the figure, four exemplary cut sets including associated deduction paths from layer 2 are given. Each cut set consists of up to two independent units that are faulty. One cut set belongs to an externally supplied component. For MOBILE, that could be one of the angle sensors at an axle. This component is not detailed to the lowest layer. To support the processing of the derived data, the algorithm has to ensure that each cut set is unique and that dual-fault based cut sets are pairwise disjoint with single-point fault based ones:

$$\begin{aligned} C^l_{i_{\{s/d\}}} &\neq C^g_{j_{\{s/d\}}} && \forall \quad i \neq j \wedge l \neq g, \\ C^l_{i_s} \cap C^g_{j_d} &= \emptyset && \forall \quad \text{valid combinations of } i, j, l, g. \end{aligned} \tag{6.2}$$

l and g reference hierarchical layers, and i and j denote the cut set counter.

Result: cut sets

After completing the above two sub-steps of step 7, a list of cut sets is available for each critical failure scenario identified at the "vehicle layer". According to the best knowledge of the developer performing the safety analysis, the combination of these cut sets forms a minimal cut set for the overall system. The only limitation is that cut sets consisting of a combination of more than two faults are neglected.

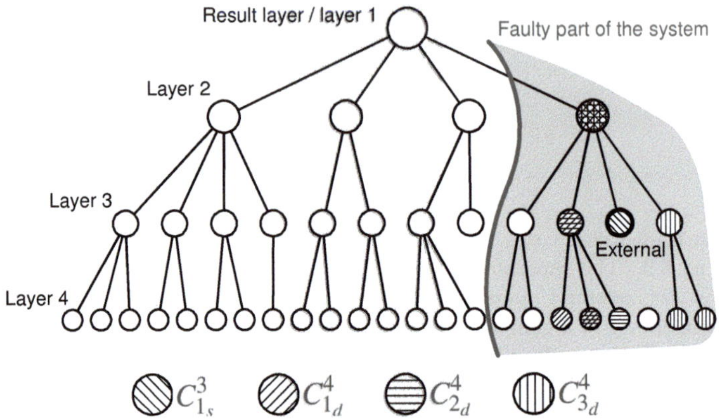

Figure 6.10.: Cut sets for the faulty part of an example system (note: for MOBILE this would be a glance at the overall hierarchy of virtual systems)

6.2.8. Step 8: Derive Top-Level Failure Rate

Step 8 derives the failure rate of the vehicle from the failure rates of components, which are associated to the identified cut sets. Therefore, the individual failure rates have to be calculated, the mission time has to be taken into account, and the emergency operation interval has to be considered. The following sub-steps result:

1. *System failure due to one single-point fault based cut set:* To start with, the failure rate due to one single-point fault based cut set is determined. In this case, a system failure represents a single Markov transition. Accordingly, the failure probability per mission can be derived from the failure rate λ_i^l, which is associated to the i-th single-point fault based cut set on layer l and the mission time T_M:

$$P(C^l_{i_s}) = 1 - e^{-\lambda_i^l \cdot T_M} \approx \lambda_i^l \cdot T_M \text{ for small lambdas.} \tag{6.3}$$

2. *System failure due to one dual-fault based cut set:* For the calculation of the failure rates due to dual-faults, the approach presented by Sieglin [2009] is adopted. Sieglin [2009] derives a formula to calculate the failure probability of a power supply system consisting of two independent units with failure rates λ_1 and λ_2 and a diagnostic unit. The structure of this duplex system is given in Fig. 6.11. The failure probability of the system p_{fail} due to a failure of both independent units can then be calculated as:

$$p_{\text{fail}} = 2 \cdot \lambda_1 \cdot \lambda_2 \cdot (T_M \cdot T_{SS} - \frac{1}{2} \cdot T_{SS}^2), \tag{6.4}$$

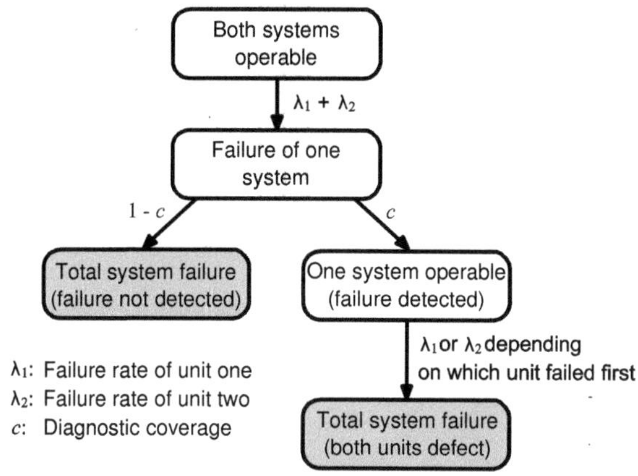

Figure 6.11.: State transitions in case of failure for a duplex system with diagnostic unit; figure similar to Sieglin [2009].

with T_{SS} being the emergency operation interval. The formula can only be applied if three assumptions are valid. (a) At mission start, all systems have to be in the "ok" state. This can be ensured by a detailed self check. (b) The exponential function $f(\Delta t) = 1 - e^{-\lambda \Delta t}$ can be approximated by the linear function $f(\Delta t) = \lambda \Delta t$ for small products of failure rate λ and time interval Δt. And (c), the failure rates do not change depending on the order of occurrence of the failures. For the hierarchical approach the assumptions (a) and (b) are valid. Assumption (c) is mostly fulfilled due to the definition of the independent virtual systems. If this assumption is not valid, the developer can easily introduce different failure rates, because the two associated failure scenarios are treated independently during the system analysis. Thus, the formula is adapted:

$$P(C^l_{i_d}) = 2 \cdot \lambda^l_{i,1} \cdot \lambda^l_{i,2} \cdot (T_M \cdot T_{SS} - \frac{1}{2} \cdot T^2_{SS}). \tag{6.5}$$

$\lambda^l_{i,1}$ and $\lambda^l_{i,2}$ represent the failure rates associated to the first and second fault forming the dual-fault based cut set with counter value i on the hierarchical layer l. The factor two ensures that both scenarios with one of the units failing before the other are included. If the failure rates change for these two cases, the evaluation algorithm accordingly neglects the factor two and treats both cases independently. Thus, the presented approach provides intrinsic means to handle such timely effects. Other approaches have to rely on special extensions, such as those introduced by Mahmud et al. [2010] or Walker and Papadopoulos [2009].

3. *Combining all failure probabilities throughout all hierarchical layers:* To combine the failure probabilities throughout all hierarchical layers, a bottom up approach is followed. Starting with the lowest layer, the failure probabilities are combined. Then, the process is repeated on the next highest layer for the cut sets that are so far untreated. The combined failure probability due to multiple cut sets at layer l is calculated as done in reliability engineering [Verma and Ajit, 2010, p. 22]. Transferred to the notion of the hierarchical approach, the following formula results:

$$\begin{aligned} P(C^l) =& (+1) \cdot \sum_{i=0}^{I} P(C^l_{i_{\{s/d\}}}) \\ & (-1) \cdot \sum_{i=0}^{i=I-1} \sum_{j=i+1}^{j=I} P(C^l_{i_{\{s/d\}}} \cap C^l_{j_{\{s/d\}}}) & \} \ \mathrm{OM}(\lambda^3) \approx 0 \\ & (+1) \cdot \sum_{i=0}^{i=I-2} \sum_{j=i+1}^{j=I-1} \sum_{k=j+1}^{k=I} P(C^l_{i_{\{s/d\}}} \cap C^l_{j_{\{s/d\}}} \cap C^l_{k_{\{s/d\}}}) & \} \ \mathrm{OM}(\lambda^4) \approx 0 \\ & \dots \\ & (-1)^{I+1} \cdot P(C^l_{1_{\{s/d\}}} \cap C^l_{2_{\{s/d\}}} \cap \dots \cap C^l_{I_{\{s/d\}}}) & \} \ \mathrm{OM}(\lambda^{I+1}) \approx 0. \end{aligned} \tag{6.6}$$

I denotes the number of cut sets within the hierarchical layer, j and k are helping variables that are defined based on i. $\mathrm{OM}(\cdot)$ qualitatively represents the order of magnitude of the terms relative to a typical failure rate λ. The orders of magnitude are derived from Equ. 6.3, Equ. 6.5, and the assumptions given in Equ. 6.2 that cuts between a single- and any dual-point fault based cut set do not exist. As a result, only dual-fault based cut sets have to be considered in line two and the following lines of Equ. 6.6. The cuts are calculated based on the conditional probabilities resulting in the given order of magnitude in terms of λ. For example, the probability of a cut between two cut sets with identifiers i and j of the same hierarchical layer with one common fault within each cut set is determined as:

$$P(C^l_{i_d} \cap C^l_{j_d}) = P(C^l_{i_d}) \cdot P(C^l_{j_d} | C^l_{i_d}) \propto \lambda^l_{i,1} \cdot \lambda^l_{i,2} \cdot \lambda^l_{j,2} \quad \} \ \mathrm{OM}(\lambda^3). \tag{6.7}$$

The faulty unit, which is part of both cut sets, is associated with the failure rates $\lambda^l_{i,1}$ and $\lambda^l_{j,1}$, respectively. Based on the assumption of small lambdas, all terms with an OM higher than λ^2 are neglected.

To conclude step 8, the failure rate F of the vehicle per mission results by adding up all $P(C^l)$ derived for each hierarchical layer:

$$F = \sum_{l=1}^{l=L} P(C^l). \tag{6.8}$$

Thus, compliance with required failure rates, such as those derived from the ASIL classification but including software, can be verified. Additionally, aging effects can be considered by an iterative execution of the outlined calculations with modified failure rates.

6.2.9. Step 9: Derive Diagnostic Coverage

Finally, the diagnostic coverage (DC) at "vehicle level" can be estimated based on the results of the hierarchical approach. As defined for an FMEDA, the diagnostic coverage is calculated as [Löw et al., 2010]:

$$DC = \frac{\lambda_{dd}}{\lambda_{dd} + \lambda_{du}}. \tag{6.9}$$

λ_{dd} and λ_{du} denote the cumulative probabilities of occurrence of any dangerous failure that is detected (dd) or remains undetected (du). This approach is now transferred to "vehicle level". In this process, it is important to note that the examined highly integrated system does not have one separate diagnostic unit, such as the one introduced by Sieglin [2009] and indicated in Fig. 6.11. The functionality of the diagnostic unit is distributed throughout the network. Furthermore, if functional redundancies between different actuators are considered, it is inevitable that the classical perception of the diagnostic unit will have to be modified towards a globally operating system, which includes the knowledge about vehicle dynamics and the vehicle network.

Calculation process

The basic idea behind the calculation is as follows: due to the targeted abstraction based on the generalized failure states, the hierarchical approach only regards dangerous faults. Furthermore, it is assumed that all safety critical functions are executed in at least two independent ways, because otherwise the high safety requirements cannot be met. Thus, any functional unit can be seen as being structured similar to Fig. 6.11. Any single-point/residual failure leading to an overall system failure then must be caused by a failure of the diagnostic system and must contribute to λ_{du}. As a result, the diagnostic coverage DC for a system that only has to tolerate one independent fault can be calculated according to Equ. 6.9 by relating faults that lead to an immediate failure of the system (λ_{du}) and faults that enable the vehicle to transition into its safe state (λ_{dd}). This requires that the local diagnostic performance is included as part of the local failure rate fed into the system. The diagnostic coverage then indicates the performance of the distributed diagnostic algorithms, and thus signals whether the failure rate of the vehicle is driven by the high failure rates of individual units or the missing quality of the diagnostic algorithms. On demand, automatic hints indicate the local diagnostic units with the lowest performance. Then, an expert for the local system can derive appropriate solutions.

6.3. Critique of the Hierarchical Approach

Fields of application

This section revisits the fields of application, novel contributions, and limitations of the hierarchical approach. The need to evaluate highly integrated control systems for Drive-by-Wire vehicles that feature functional redundancies in terms of functional safety motivated the development of the hierarchical approach. Judging from the promising research results in the field of vehicle dynamics, functional redundancies flanked by a suitable degradation concept may hugely support the cost-effective implementation of these highly safety critical EE systems. Still, also in series vehicles, increasingly more integrated control systems for critical functions, such as propulsion and braking, are introduced that might capitalize on such considerations. Appropriate methods for these safety evaluations are rarely found. Thus, the hierarchical approach aims to close this growing gap.

Contributions

Several key aspects distinguish the hierarchical approach when compared to existing works with a similar holistic perception of the vehicle:

- The hierarchical approach presents a *targeted way to conduct the system safety evaluation*. In the process, the front loading of knowledge about dependencies and critical states by using the introduced virtual systems and generalized failure states ensures that only relevant components and faults are considered in the evaluation process. This goal-oriented process facilitates that a high level of abstraction is achieved while still maintaining mathematical linkage for quantitative evaluation and proper documentation. Thus, in particular the effort to derive quantitative failure rates can be reduced compared to other approaches, such as the one by Papadopoulos et al. [2001].
- The hierarchical approach leads to an already tailored *structure function*[12] of the examined system that neglects unnecessary components. The structure function could be visualized in different ways or be reused for further efficient analysis of the system as shown by Adachi et al. [2011], Herath et al. [2007], Rehage et al. [2005], and Abele [2012]. In this work, the structure function is implicitly contained in the Excel dependency tables derived for the virtual systems and no mean for automatic visualization has been implemented so far.
- The hierarchical approach makes it possible to highlight the components with the highest impact on failure rates but also *indicates the contribution to functional safety* on each hierarchical layer. Thus, the approach supports system level thinking and encourages failure handling on all hierarchical layers.
- The hierarchical approach provides a failure rate for the overall system taking into account a configurable *emergency operation interval* and *failure rate estimates of dedicated software* functions. Additionally, the *diagnostic coverage* at "vehicle level" is approximated based on single- and dual-fault based cut sets and their probabilities of failure.

[12]The structure function defines the "dependence of the system state on the state of its components" [Gertsbakh, 2000, p. 1].

- Compared to the design approach by Abele [2012], which was applied in a series development project, the approach described in this section features several similarities. It follows a top-down approach, maintains interconnections among hierarchical layers and therefore supports the local execution of FMEAs, FTAs, the deriving of a structure function, and allows to re-use components. Still, considering cross-domain dependencies and the introduced generalized failure states for virtual systems distinguish the hierarchical approach from the approach of Abele [2012]. These aspects contribute to tailor the safety analysis and to exploit functional redundancies, which to date have not been considered in existing approaches.

Limitations: The hierarchical approach contributes to evaluating a system in terms of functional safety. Naturally, there are also clear limitations of applicability.

- As already pointed out, the hierarchical approach is neither a process model nor a method that primarily supports the product development process. The hierarchical approach supports the evaluation of an already drafted vehicle architecture. Iterative application may also support system development.
- Additionally, the tailoring of the hierarchical approach with regard to functional safety evaluation reduces work effort but may also be inappropriate for the extended investigation of a system, e.g., for a full analysis of the system reliability. If the hierarchical approach is accordingly extended, work effort approaches the one of the already existing methods. The other way round, if generalization during the hierarchical approach is overdone to further decrease work effort, safety estimates become more pessimistic. Still, the results from common methods, such as FMEA, support the developer to set the abstraction level appropriately by identifying the relevant failure modes.
- The hierarchical approach focuses on the quantitative evaluation of failure rates but does not evaluate the fulfillment of process requirements given in ISO 26262. Process requirements derived from ASIL levels are a vital aspect for the safety evaluation [Palin et al., 2011] and significantly contribute to development costs. The hierarchical approach could contribute to reduce process requirements by identifying local redundancies that facilitate ASIL decomposition. But, the treatment of functional redundancies across different types of actuators in terms of ASIL classification is so far a mainly uninvestigated challenge for research and development.
- Until now, the hierarchical approach only focuses on the basic vehicle control functions. Other functions provided by the human machine interface are not considered. Still, this contribution holds the view that none of these aspects is relevant if the driver can no longer control the main actuators of the vehicle. Thus, the vehicle control system forms the basis for any other applications and should be treated separately. This perception is backed by the generic safety life cycle for intelligent transport systems outlined by Carsten and Nilsson [2001].

- The quality of the results generated with the hierarchical approach depends on the expertise of the developers performing the analysis. The method only supports the structured gathering of information and the following quantitative evaluation. This restriction due to the dependence on the developer applies to all comparable methods for a-priori safety analysis of a system that have been found in the literature.

Outlook

The hierarchical approach forms a first step towards the structured and tailored analysis of functional safety for novel integrated EE systems. The information gathered by the approach could be further exploited in multiple applications. For example, the resulting structural information could be used for the optimized allocation of failure rates or failure detection mechanisms as presented by Herath et al. [2007] and thus guide product development. Also, the structured and tailored safety information on the vehicle could be exploited for an online self-representation system for the vehicle as will be introduced in Sec. 8.2. The vehicle may then include this information about itself in automated decision making processes in case of failures both in the field of automated driving and for automated control allocation to actuators. Finally, the missing quantification of failure rates of control or environmental perception algorithms in the fields of vehicle dynamics and automated driving poses a huge challenge for future research. The hierarchical approach showed the need to investigate these algorithms quantitatively if functional redundancies shall be exploited to prove functional safety. However, no such approaches are available yet, which leaves this task to future research.

Part IV: Enabling Functional Safety Efficiently

Achieving a required level of functional safety in a Drive-by-Wire vehicle efficiently is challenging and requires dedicated mechanisms, which extend the EE architecture proposed in Part III. Therefore, part IV introduces specialized mechanisms that contribute to address this challenge. Both short (tactical) and long-term (strategic) mechanisms are proposed.

7

Tactical Safety Measures

"Tactics mean doing what you can with what you have."

Saul Alinsky

"Zebra tactics" for MOBILE

If an animal gets wounded, either by accident or while being hunted by a predator, the only chance for it to survive is to adapt to the new circumstances and make the best out of what it still can do to remain "operable". Taking the example of a zebra fleeing a lion and getting injured at a leg, it will go on running in the best possible way relying on the remaining three legs for an "emergency operation interval". And, if there was no lion, it would most likely come to a safe stop. Basically, it "tactically" exploits the available functional redundancies in its actuators, the legs, to maintain a basic movement. In doing so, the coordination of the open-loop motoric programs controlling the movement of the legs is modified based on sensory feedback [Meiner and Schnabel, 1987; Dickinson et al., 2000]. Barely any highly developed animal ever has "back-up" extremities that are providing a classical redundancy in case of injuries – unlike some organs, such as kidneys or left and right lung that are available redundantly [Cherry, 2000]. As a result, the zebra – just as most other animals – is forced to rely on functional redundancy and thus it is obviously vital how well it adapts to the new situation.

Taking this approach from nature as a motivation, the functional safety of MOBILE is strongly based on a degradation concept: it mainly relies on dynamic reconfiguration of the system in case of partial system failures. Then, the key issues are to monitor the status of the system including the electronics and the actuators, to detect occurring faults, and, if needed, to reconfigure the system to prevent critical failures. Analogously to the zebra, these algorithms operate tactically[1] by reacting quickly to an unforeseen immanent threat or change of the system state. The performance of these algorithms significantly contributes to the

[1]In terms of nomenclature, this thesis refers to tactics and strategies depending on whether the associated actions or considerations aim at a short term perspective or at long-term goals with a focus on the overall system. This agrees with the usage of the terms by Siedersberger [2003] to describe the operation of a highly automated vehicle: strategic measures consider the overall mission (extended in this work: and the lifetime of the vehicle). Tactics react to the current (traffic) situation. Other works in the field of automated driving, such as the one of Schneider [2010], refer to similar classifications. Also in human decision making, plans [Rasmussen, 1983,

P.J. Bergmiller, *Towards Functional Safety in Drive-by-Wire Vehicles*,
DOI 10.1007/978-3-319-17485-3_7

quality of the safety concept, and thus ensures the controllability of the vehicle after a partial system failure. At the same time, these tactical measures are vital to reduce the number of required hardware redundancies in the vehicle. The following chapter first outlines the online monitoring algorithm applied in the vehicle. Then, an outlook on the ongoing research in the field of vehicle control to generate the desired vehicle handling even in case of actuator failures is provided.

7.1. Probabilistic Fault Detection and Handling (PFDH)[2]

For the safety concept associated with the architecture introduced in Cha. 5, the fault tolerant units controlling each axle are vital. As mentioned, each unit consists of two redundant controllers: a primary and a secondary node (Fig. 5.9). The primary node serves as a development platform with little restrictions due to the functional safety requirements. The secondary node performs the online monitoring and provides basic driving functionality in case of failure (Fig. 5.10). The newly developed software on the primary node is highly prone to errors, while at the same time details of the software are mostly unknown ("black box"). This forms a challenging task for the monitoring system executed by the secondary node. Still, several sources of information are available to the secondary node: models of vehicle dynamics, predefined communication and execution schedules, online diagnostics of all nodes in the network, and the in- and output signals of the primary node. Figure 7.1 summarizes the relation between the primary and the secondary node within a fault tolerant unit. To additionally ensure a fail-silent behavior for itself, the secondary node is assisted by the power control units.

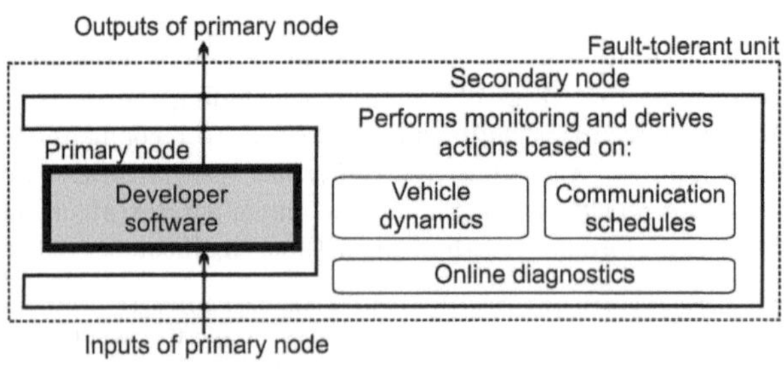

Figure 7.1.: Black-box approach within a fault-tolerant unit

p. 259] equaling strategies [Maurer, 2000, p. 29] are associated to the top-level knowledge based behavior. Still, borders between the two aspects may blur depending on the examined scenario.

[2]Parts of this section have been pre-published by the author in Bergmiller et al. [2011b] and Bergmiller and Maurer [2011].

Table 7.1.: Requirements on PFDH derived from top-level requirements (Tab. 4.1)

R1	**Mechanical and electronic modularity**
R1.P1[a]	PFDH shall enable the flexibility of the EE system by supporting easy exchange of the primary controllers or the software running on these.
R2	**"Open-source" vehicle**
R2.P1	When using the experimental vehicle as a test bed, the developer should not primarily have to worry about the functional safety.
R2.P2	PFDH has to be easy to configure and to adapt to novel vehicle configurations.
R3	**Functional safety**
R3.P1	PFDH has to support a degradation concept to maintain basic driving functionality even in case of partial system failures. The concept should support measures for self-healing of the system.
R3.P2	All decisions made by PFDH have to be traceable, and unexpected behavior in uncertain conditions has to be avoided.
R3.P3	Robust decision making is important for application in real vehicles and to reduce dependence on precise quantitative data.
R3.P4	PFDH has to support low execution times on microcontrollers.
R4	**Limited degree of hardware redundancies**
R4.P1	PFDH has to take over the role of the diagnostic unit in a system with dynamic redundancy (1oo2D)[b] to economize a third component.

[a]R1.P1 stands for the first requirement on PFDH derived from the top-level requirement R1.
[b]Explanation will be given in Sec. 7.1.1.

Requirements

In detail, several requirements on the proposed Probabilistic Fault Detection and Handling algorithm (PFDH), which is executed by the secondary node, can be derived from the requirements posed on the overall vehicle MOBILE (Tab. 4.1). PFDH mainly aims to fulfill the safety goals while at the same time not violating but supporting the modularity and usability of the vehicle. Table 7.1 summarizes these requirements.

Solution strategy

To fulfill the requirements, the introduced black-box set-up supports the flexible modification of the vehicle software without having to worry about safety extensively (→ R2.P1, → R1.P1). For monitoring of the black box, PFDH relies on multiple simple criteria rather than few computationally intense factors → R3.P4. These inputs are processed following a probabilistic approach → R3.P2, which supports real-time execution on a microcontroller → R3.P4. If an action has to be taken, PFDH chooses from a list of available options. Depending on the detected fault, a stepwise escalation of actions is performed to minimize impact on the remaining system [→ R3.P1], or trigger self-healing measures. PFDH gains robustness in decision making by comparing different action alternatives for different network nodes [→ R3.P3]. This way, no vague absolute thresholds with a huge impact on the system performance have to be defined by the expert configuring PFDH [→ R2.P2]. In cooperation with the power supply controllers, PFDH makes it possible to limit the number of independent ECUs inside an axle to two [→ R4.P1]. It supports the reuse of available computational resources for diagnostics if special mechanisms for the safe sharing of resources are implemented on the secondary node (see also Sec. 5.1).

Core contributions

The PFDH algorithm designed according to the above outlined strategy features a combination of several key advantages that is not provided by other systems. The system supports (a) easy integration of the expert knowledge and generates (b) robust decisions based on a comparative approach. At the same time, the system is (c) mostly insusceptible to sporadic faults but generates fast decisions in case of clear error conditions. The decisions made by the system are (d) traceable and predictable for so far not considered situations. The support for (e) the degradation concept for MOBILE, which favors targeted actions, contributes significantly to the safety concept. Therefore, a simple measure for the costs of an action is proposed. (f) The monitoring algorithms themselves are considered as a part of the system that is also susceptible to faults. This is barely done in related research projects. Finally, the overall system requires (g) little computational resources if implemented appropriately, and thus it is executable on microcontrollers.

7.1.1. Related Work

Related work for PFDH can be found in multiple fields of application that make use of special system structures and algorithms for decision making and online reconfiguration. Still, all approaches have to deal with common challenges. Accordingly, the related work section is structured. First, the set-up of PFDH (1) is put into relation with the typical redundancy structures. Next, ways to derive the symptoms[3] to monitor a technical system are outlined (2). Based on these symptoms, the decision units (3), which are introduced in literature to determine appropriate actions are investigated. Finally, PFDH is related to the existing approaches (4).

(1) System structures

In the literature, several different redundancy schemes are discussed to set-up systems that are classified as fail-operational, fail-safe, or fail-silent. For the auto-

[3] A symptom is regarded as "any phenomenon or circumstance accompanying something and regarded as evidence of its existence" [Collins, 2013]. In the medical field, the expression "symptom" is used in a way similar to its usage in common language. For a technical system, this thesis assumes that a symptom indicates the presence of an "error" that may or may not lead to a system failure. According to ISO 26262, an error is defined as "a discrepancy between a computed, observed or measured value or condition, and the true, specified or theoretically correct value or condition" [ISO 26262-1, 2011, p. 7], which can result from a fault or an operation under unforeseen conditions.

motive domain, fail-operational refers to a system that "stays operational after one failure" [Isermann et al., 2002, p. 69], fail-safe characterizes a system that transitions into a safe state "after one or several failure(s)" [Isermann et al., 2002, p. 69], and the fail-silent behavior demands that "the component exhibits quiet behavior externally and therefore does not wrongly influence other components" [Isermann et al., 2002, p. 69]. For the intended field of application of PFDH, especially the fail-operational systems are of interest, because the controllability has to be maintained. To achieve such a behavior, hardware and analytical redundancy can be implemented (compare Sec. 5.1). For redundant units, diverse redundancy refers to units that feature different structures, rely on different measurement principles, or run software implemented by different developers. As a result, the systems are less susceptible to common cause failures [Isermann, 2008, p. 565]. To achieve a fail-operational behavior in a system with static redundancy, meaning with no monitoring and reconfiguration unit, at least three components providing the same functionality or value are required [Isermann et al., 2002, p. 70]. If a diagnostic and action derivation unit is added, the minimal number of required devices for redundancy can be lowered to two units. Kirrmann and Großpietsch [2002] associate the nomenclature "1oo2D" to such a structure. 1oo2 means that "one out of two" devices needs to be operational, and the D indicates the presence of a diagnostic unit. As outlined in Sec. 5.1, a 1oo2D structure is acceptable for a system that has to provide only a short emergency operation time to achieve a safe state. This applies to MOBILE.

(2) Deriving symptoms

Assuming a 1oo2D structure, a vital task for the monitoring unit is to generate appropriate symptoms that indicate the health state of the system components. Therefore, different approaches exist. Isermann et al. [2002] distinguish three types of symptoms depending on their origin: (1) limit value checking and plausibility checks monitor the value ranges of signals. (2) The signal model based methods investigate periodic or stochastic signals, while (3) the process model based methods correlate two or more signals via the known dependencies between the signals and derive the appropriate symptoms. The latter group of methods has been analyzed since the 1970s and seems promising, because it achieves a good diagnostic performance while limiting the amount of additional hardware units for safety purposes [Schwall, 2005]. Still, complex model based methods require more computational power [Isermann and Beck, 2011]. Some examples for process model based methods are presented by Hermans and Zarrop [1996] and the already referenced research groups of Isermann [Isermann and Beck, 2011; Isermann, 2008; Isermann et al., 2002; Muenchhof et al., 2009] and Gerdes [Gadda et al., 2007, 2004; Schwall, 2005].

(3) Determining actions

Starting from the generated symptoms, an action derivation unit has to determine whether a system failure is present, and which action has to be taken. For MOBILE, an additional emphasis is put on selecting targeted actions that focus specific faults and therefore limit the impact on the remaining system. In the literature, several action derivation strategies are presented, which range from intuitive rule-based approaches to methods from the field of artificial intelligence that make use of in-field measurements from a fleet of systems. Some exemplary systems with closest relation to the PFDH approach are introduced.

Cause-symptom correlation

A basic issue when deriving actions based on symptoms has long been known in the medical field and is addressed by the research group of Isermann [Isermann, 2008; Muenchhof et al., 2009] for electromechanical systems: one symptom frequently does not uniquely indicate a certain error and its underlying fault. Often, different faults can only be distinguished by comparing symptom patterns which makes the cause identification challenging. Therefore, Muenchhof et al. [2009] outline a fault-symptom correlation matrix for electro-mechanical actuators. These matrices make it possible to reversely derive the faults given in the rows of the matrix from the detected symptoms associated to columns. Basically, the tables can be understood as "If-then-Else" systems written as matrices, which make it easier and more intuitive for an expert to enter knowledge about the system structure. The clear separation between symptoms and responsible faults is also desirable for MOBILE to support the stepwise system analysis and the intuitive configuration of the action derivation system. For the examples introduced by Muenchhof et al. [2009], the action derivation based on the identified faults is not detailed. Especially, the reaction of the monitoring system to uncertain error conditions and the handling of multiple available actions are not considered. These influences are captured by an approach proposed by Schwall [2005] based on Bayesian Networks. Schwall [2005] also starts with deriving possible causes of detected symptoms but then continues to determine possible actions based on the identified cause probabilities and predefined costs of an action.

Fuzzy Logic

Fuzzy Logic based systems constitute another frequently applied approach to monitor systems and derive actions based on symptom patterns. This approach stems from the field of artificial intelligence and was already introduced in the 1960s. Fuzzy Logic is especially suitable for applications where inputs from experts have to be acquired to form the decision making system. If a suitable fuzzy set and according membership functions for the current application have been defined, the decision making system can be set up intuitively by the developer [Kruse et al., 2011, 1994; Zadeh, 1965]. Khan [2007] presents a Fuzzy Logic based diagnostic system for a turboprop engine to detect anomalies and predict the future system degradation to schedule maintenance. Shen et al. [2012] detail a fault tolerant control system for near space vehicle attitude dynamics that integrates Fuzzy Logic for decision making in case of failures. Further examples using Fuzzy Logic for the decision making are referenced by Shen et al. [2012] for technical applications and by Geman [2011] in the medical field. In general, fuzzy approaches can become impracticable if a large number of symptoms is available that has to be associated to multiple possible actions. Then, a matrix based approach as introduced by the group of Isermann is better suited to handle complexity.

Bayesian Networks

A Bayesian Network (also called belief network or influence network [Pearl, 1986]) is a powerful mechanism for action derivation, but it is frequently also computationally intense. Bayesian Networks are especially applicable if "the knowledge [to solve a problem] is difficult to acquire or is based on rules that can only be learned through experience" [Castillo et al., 1997, p.8]. Bayesian Networks became a viable solution for the real-world scenarios when suitable algorithms to solve the networks were introduced by Pearl [1986, 1988]. Only the approach by Pearl [1986] made it possible to deal with the exponentially increasing complexity associated with the calculation of the joint probabilities of multiple input signals. Since then, the fields of application for Bayesian Networks have been growing [Murphy, 2002]. An important strength of Bayesian Networks is that a human expert can configure it such that each node in the network is associated a meaning in terms of the application, e.g., a node can represent the failure probability of a dedicated component. This makes the definition of the conditional probabilities listed in the conditional probability tables (for discrete nodes) associated to each node more intuitive and facilitates plausibility checks of an existing Bayesian Network. Apart from manual configuration by experts, Bayesian Networks can also be generated or parametrized based on existing measurements by applying appropriate learning algorithms. Basically, algorithms to learn the conditional probabilities and algorithms to derive the graph structure are distinguished [Koski and Noble, 2009]. Overall, Bayesian Networks are a powerful tool for decision making. Again, the medical sector exploits the capabilities of Bayesian Networks to derive the causes of symptoms shown by a patient [Geman, 2011; Spiegelhalter et al., 1993]. In the technical field, Bayesian Networks are applied in multiple applications for online monitoring [Camci and Chinnam, 2005; Murphy, 2002; Schneider, 2010; Schwall and Gerdes, 2002]. Dynamic Bayesian Networks integrate time aspects to link Bayesian Networks, which represent the system states at a given time slice, in the manner of a Markov-Chain. For further information on Bayesian Networks in General, refer, e.g., to Koski and Noble [2009], Bishop [2006], and Murphy [2002], who provide the basics of Bayesian Networks, a relation to other machine learning approaches, and a detailed discussion of fields of application. The toolbox implemented by Murphy [2001] is used in this work for later comparison between PFDH and Bayesian Networks (Sec. 7.1.3). As a key issue when using Bayesian Networks, it has to be defined when a failure probability is "high-enough" to take an action. The probability based threshold is easier to understand for the developer than other criteria directly linked to signals [Schwall, 2005, p. 15], but still it requires an absolute threshold, which can be hard to set [Emami-Naeini et al., 1988].

Neural Networks

To complete the presented outline of available decision making mechanisms, Neural Networks shall be referenced. The development of Neural Networks started with the investigation of single neurons in 1943. Still, it took until the 1980s before Neural Networks were really used in applications. At that time, the usability increased significantly with the development of novel learning algorithms that allowed to train multi-layer Neuronal Networks [Müller, 2011]. These networks are able to approximate any function for which the Riemann integral can be calculated [Kruse et al., 2011, p. 50]. Müller [2011] gives a more detailed historical view on the development of Neural Networks and references the relevant literature. Theoretical details can be found in Kruse et al. [2011]. This paragraph points out the downsides of Neural Networks, which make them unsuitable for the proposed system. A key disadvantage of Neural Networks is the lack of interpretability. Other than in

Bayesian Networks, the individual neurons (nodes) do not have a dedicated meaning from the application point of view, which hinders plausibility checks by the developer. For the same reason, a Neural Network can usually only be generated based on training data. This data is frequently unavailable or incomplete – especially for newly developed systems. For so far untested conditions, the behavior of a Neural Network is unknown and may cause severe failures in decision making. In general, Neural Networks can hugely contribute to non-safety-critical decision making where maybe even a human expert is involved for online plausibility checking of the final results. Numerous applications can be found in the medical field [Kondo, 2011; Saito and Nakano, 1988; Walczak, 2005] or for non-safety critical applications in the field of vehicle electronics [Müller et al., 2011]. The mentioned downsides of Neural Networks for safety-critical applications apply to most other machine learning based methods [Hastie et al., 2009], unless special measures are taken [Bergmiller et al., 2008], which are not suitable for all applications.

(4) Relating PFDH

Concluding the related work section, PFDH is briefly put into relation to the outlined research approaches. As stated, PFDH has to derive traceable decisions for safety reasons → R3.P2 and has to be easily understandable and configurable by the developer → R2.P2. An unexpected behavior of the system has to be avoided. For this reason, most machine learning based approaches including approaches based on Neural Networks are unsuitable. Nevertheless, the approach to learn from already gathered data to improve the system performance is useful and has to be examined with regard to optional integration into PFDH. Short execution times of PFDH ($<$ 4ms) on a platform with limited computational resources contribute to a quick fault detection → R3.P4 but also limit the applicability of the complex decision making systems, such as systems that rely on Bayesian Networks and special solving algorithms. Still, timely effects, such as considered by Schwall and Gerdes [2002] by using Dynamic Bayesian Networks, shall be considered by PFDH. Closest to the requirements in the project is the stepwise cause derivation proposed by the research group of Isermann. This system derives decisions traceable, quickly, and it is easy to configure for the user. For application in MOBILE, the main deficits of this approach are the missing ways to handle uncertainty when making decisions, to include timing effects, to define decision thresholds, and to select appropriate actions while taking into account the costs of the action (→ R3.P1, → R3.P3). Some of these aspects are already considered by the approach of Schwall [2005].

7.1.2. The PFDH Approach

To overcome the deficits of the outlined approaches with respect to the needs in the MOBILE project, PFDH is structured as outlined in Fig. 7.2. Three main steps are taken: the monitoring unit generates probabilities of error ❶ for each input according to a predefined monitoring strategy. These probabilities of error are merged into S groups based on common origin or content, which consist of N_s signals each ❷. For example, all inputs indicating timing problems are grouped and form a symptom. An error probability is calculated for each symptom and

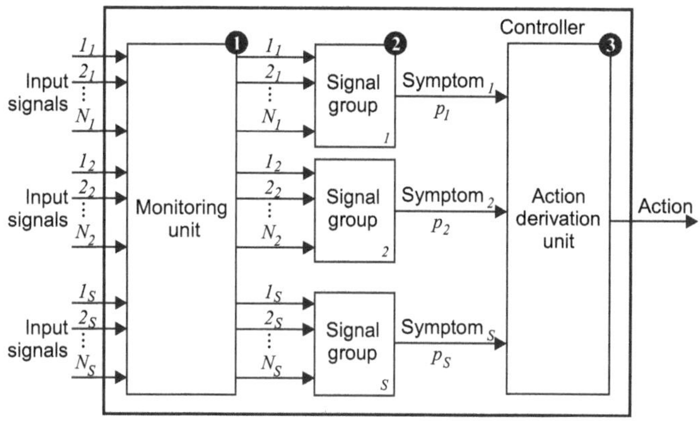

Figure 7.2.: Schematic of the Probabilistic Fault Detection and Handling algorithm

forwarded to the action derivation unit. The action derivation unit takes these probabilities as a basis to determine the action ❸ with the highest probability of success to address the existing faults at lowest costs. In the following, the given steps are detailed. To start with, the monitoring strategy is briefly summarized to serve as a basis for the consecutive introduction of the action derivation process. More details on the monitoring strategy will be given during evaluation of PFDH.

Monitoring Strategy

The monitoring unit provides input signals for the action derivation process. The signals are generated by the secondary node without knowing the algorithms executed by the primary node (black-box approach, Fig. 7.1). Three sources of information are exploited by the secondary node:

- The send rate and content of signals transmitted by the primary node are monitored to detect missing signals and a faulty chronological order or content of signals.
- Input and output signals of the primary node are related to distinguish between failures due to faulty inputs and faults within the primary node.
- Self-diagnostic functions provide information on the current health state of all nodes in the network. For this, the timely synchronization among the nodes supported by FlexRay is exploited to perform majority votes on the alive state of each node.

According to these sources of information, the application specific symptoms are defined that will be introduced in Sec. 7.1.3.

Symptom Error Probabilities

Based on the results of the monitoring unit, the probability of a symptom being detected, p_s, is set to:

$$p_s = max\left(\frac{\sum_{n=1}^{n=N_s}(h_{s,n} \cdot p_{s_n})}{\sum_{n=1}^{n=N_s}(h_{s,n})}, E_s\right), \quad s \in \{1..S\}, \ n \in \{1..N_s\}, \ h_{s,n} \in \mathbb{R}^+, \quad (7.1)$$

where h_{s_n} denotes the weight of the n-th signal within the signal group s with N_s members and p_{s_n} is the error probability of the n-th signal. E_s is an error probability for the complete signal group set by the self-diagnostic functions.

Action Derivation

Based on the error probabilities indicated by each signal group, the action derivation process is initiated, which consists of several core steps (**A-F**).

(A) - Error perpetrator

To start with, it is determined whether the primary or secondary node caused the error. The probability that the signal processing of the secondary node for signal group s has caused the symptom is calculated based on p_s according to Bayes'theorem with PN and SN referencing the primary and secondary node:

$$\begin{aligned} &P_s(\text{"SN caused symptom } s\text{"} \cap \text{"symptom } s \text{ present"}) = \\ &\qquad p_s \cdot P_s(\text{"SN caused symptom } s\text{"} | \text{"symptom } s \text{ present"}). \end{aligned} \quad (7.2)$$

P_s("SN caused symptom s" | "symptom s present") denotes the conditional probability that the secondary node caused the symptom given a symptom is detected. This probability can be defined either based on statistical evaluations or expert knowledge and depends on the used hardware and algorithms. A simple example for this conditional probability being 0.5 is if both controllers read in a signal from CAN with the same CAN unit, the same software drivers and the same type of wire. PFDH only requires to qualitatively set this conditional probability parameter to work properly. The secondary node also calculates P_s("PN caused symptom s" $\cap$ "symptom s present") for the primary node. The events "SN caused symptom s" and "PN caused symptom s" are treated as being disjoint for single faults. If both controllers are well isolated from common sources of error like data wires or power supply units, this is a valid assumption supported by the introduced architecture. If dependencies between the two units cannot be neglected, the fault tolerant unit should be redesigned.

(B) - Signal group error probability vectors

The probabilities derived according to Equ. 7.2 for the primary and secondary node are combined into one vector for each node. For simplicity, the following formulas refer to these vectors as $\vec{e}$ with length S and do not distinguish between the two controllers. All following formulas are evaluated by the secondary node both for itself and the primary node.

(C) - Cause identification

It is necessary to identify the underlying faults that caused the detected symptoms (compare Isermann [2008] or Schwall [2005]) as a basis for action derivation.

This resembles the approach in the medical field, which also usually focuses on the treatment of the causes rather than the symptoms. Still, it is important to note that the symptoms are not exclusively triggered by the faults in the system but can also be caused by external circumstances. Therefore, PFDH neglects implausible symptom patterns that cannot be associated to known internal faults. In future, the detection rates can be further improved if external disturbances are modeled as "external faults". The possible faults of the EE system are derived based on a cause-symptom correlation matrix (**CS**). The matrix lists the possible causes (R rows) for each symptom (S columns). Each element $ce_{r,s}$ of the matrix represents a correlation between a cause/fault and a symptom and is set to 1 or 0 for an existing correlation or no correlation. Thus, the probabilities c_r for each fault are calculated (Equ. 7.3) and combined into a vector $\vec{c} = (c_1, .., c_R)^T$:

$$c_r = \frac{\sum_{s=1}^{s=S} ce_{r,s} \cdot e_s}{\sum_{s=1}^{s=S} ce_{r,s}}. \tag{7.3}$$

Equation 7.3 can only be an approximation due to the limited knowledge about the monitored system. A critical discussion is provided in Sec. A.3.

Based on $\vec{c}$, possible actions are derived. Therefore, an action-cause correlation matrix (**AC**) is introduced. This matrix is structured analogously to **CS** and provides the possible actions (G rows) to address a fault (R columns). Each element $ac_{g,r}$ of the matrix represents the probability that action g is able to treat fault r. For each action, an overall probability of success a_g is calculated:

(D) - Action derivation

$$a_g = \prod_{r=1}^{r=R} (ac_{g,r} \cdot c_r + (1 - c_r)). \tag{7.4}$$

Equation 7.4 integrates two cases: "fault not present $(1 - c_r)$", and "existing fault treated properly by action g $(ac_{g,r} \cdot c_r)$". In both cases, the system is operable after application of the action. The resulting probabilities of success for all actions are combined into a vector $\vec{a} = (a_1, .., a_G)^T$.

A cost function is introduced to avoid that always the most powerful, but also most risky action is applied. E.g., switching off one of the controllers of a fault tolerant unit addresses all faults that may exist inside the controller, but also significantly impacts the remaining system as all tasks have to be transfered to the other controller. To derive the appropriate costs, different approaches can be taken. In order to simplify the system design process, a formula to automatically calculate the costs based on the given probabilities of error and success is proposed as one possible option. This approach delivers useful results under the assumption that actions that address many causes (correlating to several high valued entries in the respective line of the **AC** matrix) have also the strongest impact on the system. E.g., a simple software reset of a component is more likely to be executed successfully than a hardware reset of the same component. If less risky actions also feature multiple high valued entries in the respective line of the **AC** matrix, this simple approach to derive costs becomes unsuitable. Alternatively, predefined costs

(E) - Action weighting

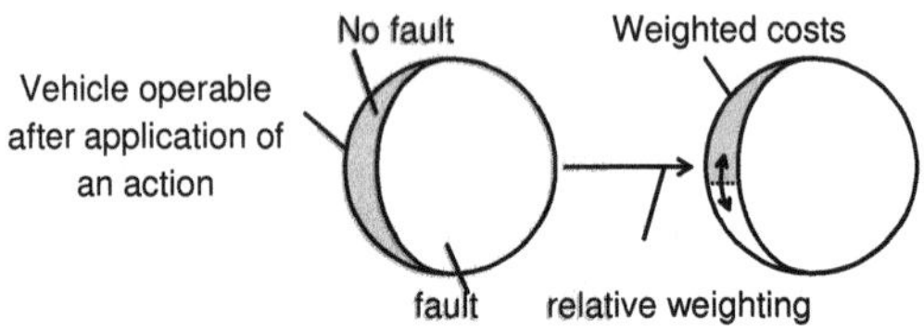

Figure 7.3.: Illustration of cost derivation in the sample space

set by the developer as done by Schwall [2005] can be used for each action. This would probably be the preferred approach as soon as the list of available actions in MOBILE has been completed, and the risk of each action can be determined based on experiences from test drives with MOBILE. For the simple approach described above, the cost of each action u_g is defined based on two influences: on the one side, the probability that the system was not faulty to start with, and thus an – possibly even risky – unnecessary action is wasted. On the other side, an additional weighting factor w_g is introduced to favor focused actions that treat an identified fault with less "collateral damage". This weighting strategy is based on the above given assumption that actions with less and smaller entries in the respective line of the **AC** matrix are also less likely to go wrong. This is a first empirical measure and valid for the set of actions implemented in the first version of PFDH on MOBILE. w_g is calculated as:

$$w_g = \frac{\sum_{r=1}^{r=R} ac_{g,r} \cdot (1 - c_r)}{\max_g(\sum_{r=1}^{r=R} ac_{g,r} \cdot (1 - c_r))}, \tag{7.5}$$

and accordingly u_g results:

$$u_g = w_g \cdot \prod_{r=1}^{r=R} (1 - c_r). \tag{7.6}$$

Figure 7.3 visualizes this cost derivation process by depicting the two mentioned steps (base costs and weighting process). The gray shaded area depicts the costs. All costs u_g are again summarized in a vector of costs $\vec{u}$. Based on $\vec{a}$ and $\vec{u}$, a vector $\vec{v} = \vec{a} - \vec{u}$ with g elements is calculated, which represent the effectiveness of each action to treat a certain fault. In each time step, the action with the highest effectiveness v_g is executed.

(F) - Time dependency

Basically, the previous step concludes the action derivation process. Still, the strong need to fix faults in a Drive-by-Wire system necessitates consideration of timely aspects to improve the performance of PFDH with regard to treatment of minor but persistent faults. Therefore, two strategies are implemented:

- The effectiveness of an action g unsuccessfully applied to treat a fault is set to zero for this fault. The success of action g is determined by checking whether an error pattern is re-detected after the action has been executed.

- The probability of a fault being present $\vec{c}^*(t_h)$ in the action derivation step at time step t_h is increased over time if the fault is re-detected to reflect the reduced trust of PFDH into the system. $\vec{c}^*(t_h)$ then replaces $\vec{c}$ in the action derivation process:

$$\vec{c}^*(t_h) = \vec{c}^*(t_{h-1}) + (1 - \vec{c}^*(t_{h-1})) \cdot \vec{c}(t_h). \tag{7.7}$$

  $\vec{c}(t_h)$ denotes the error probability vector $\vec{c}$ calculated for the current (h-th) cycle according to step **(C)** without taking history into account. Equ. 7.7 can be perceived as a simple transition in a first order Markov-Chain [Köhler, 1983] that describes the trust of PFDH into the system. Within this Markov-Chain, the previously calculated error probabilities $\vec{c}$ denote the transition probabilities, and the new probabilities $\vec{c}^*(t_h)$, which are then fed into the action evaluation process, reflect the current trust into the system with regard to the possible causes. If symptoms are autocorrelated, the failure made by assuming timely independence in Equ. 7.7 has to be assessed for the individual case. For a critical discussion of such aspects, refer to Schwall [2005]. For PFDH, this assumption holds mostly true, as discrete symptoms form the basis of the process, and the symptoms are sampled cyclically and slow compared to noise dynamics. If no cause is detected at a time step ($\vec{c}(t_h)^- = \vec{0}$), $\vec{c}^*(t_h)$ is reset to zero.

Again, this approach imitates human behavior: "When a hypothesis is confirmed with enough confidence, an attempt is made to repair the believed problem. If the repair fails, then the expert restarts the diagnostic process" [Fink and Lusth, 1987, p. 343], of course, taking into account the gained knowledge from the failed attempt. For MOBILE, it is additionally assumed that taking an action in case of uncertain fault conditions during multiple consecutive time steps is preferable to staying inactive if the fault pattern is reasonable and a dedicated fall-back layer with lower failure rates exists. If a system state with no active symptoms is detected, all changes applied over time are reset. This is based on the assumptions that the conditional probability that no symptom is generated if a detectable fault occurs is close to zero, and that a present fault is likely to be permanent.

7.1.3. Evaluation of PFDH

PFDH was evaluated in a simulation environment, on a test bench for MOBILE (see Fig. A.4 in the appendix), and with the experimental vehicle MAX (Sec. 4.1.2). Experiments with the full-scale vehicle MOBILE (Sec. 4.1.1) were not possible as the vehicle was not completed at that stage. The configuration of PFDH and the set-up of the fault tolerant units was the same for all three test environments. The special purpose of the three test benches was as follows:

- In simulation, the correct operation of the algorithms was tested and statistical evaluations of the reaction of PFDH to unexpected fault patterns and decision latencies were evaluated.

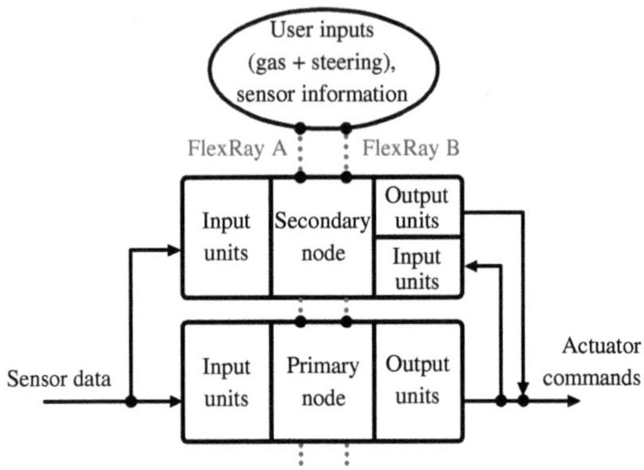

Figure 7.4.: Schematic of the structure within a fault tolerant unit

- On the test bench of MOBILE, PFDH was fed with the outputs of signal monitoring algorithms that evaluated the partially noisy signals in the vehicle. This showed that the system is real-time capable using microcontroller hardware and that PFDH operates well with non-perfect input signals.
- Onboard MAX, a first outlook on signals that take into account vehicle dynamics was made. In this set-up, two additional signals indicated the instability of the vehicle. The experiments showed that the computational power of the network nodes suffices to execute PFDH while processing more complex input signals in parallel. The additional signals enabled PFDH to distinguish between failures of the internal EE system and deviations caused by an instability of the vehicle.

In summary, the main evaluation of PFDH was performed in simulation. The outlooks on the test bench and MAX demonstrate that PFDH operates as intended on microcontroller hardware and with real measurements as inputs. The following evaluation of PFDH using an example configuration only refers to the simulation results and the test bench results, as these suffice to indicate the proper decision making of PFDH and the real-time capability.

The fault tolerant units executing PFDH during the tests are structured as shown in Fig. 7.4. The primary node takes the role of the black box (compare Fig. 7.1) and controls the actuators. Additionally, it simulates different failure scenarios, such as unintended delays, missing or wrong control commands, and faults detected by the self-monitoring system. The secondary node gathers information by measuring the in- and outputs of the primary node and reading the FlexRay bus. Both the primary and the secondary node are connected to the actuators and can take over control. It is important to note that the probability of disturbance among the nodes due to galvanic or electromagnetic coupling is strongly limited by appropriate passive protective circuits at the in- and outputs of the nodes.

Monitoring Concept and Matrix Layout

As already introduced, the PFDH based system monitors multiple simple signals rather than a few complex signals to achieve a good real-time performance. It focuses on two main fields: "determinism monitoring" and "content monitoring", which are summarized in the following.

Determinism monitoring focuses on the monitoring of the communication schedule and the internal health monitoring performed by the operating systems of all network nodes. Both aspects are enabled by the predefined and deterministic structure of the data exchange and application execution within the network. For the demonstration application, most results of the determinism monitoring are represented as "ok" or "not-ok" flags. PFDH interprets these flags for the n-th signal in a signal group as the error probabilities $p_{s_n} = 0$ ("ok") or $p_{s_n} = 1$ ("not ok"). The monitoring is done on "cycle-basis" every 4 ms. Each signal group, which forms a symptom that is fed into PFDH, consists of 2 to 10 individual input signals.

Internal health monitoring

The internal health monitoring of all network nodes contributes greatly to the fail-silent behavior of each node. For this purpose, the operating system provides several diagnostic flags which cover the most important components and tasks of the controller:

- Data acquisition, data processing, and data transmission tasks are monitored. Irregularities are detected based on violations of tasking tables.
- All controller peripherals for the data exchange and processing provide internal error diagnostics that are summarized.
- The Alive-Network-Management Vectors provide each node online access to a majority vote on the alive state of all nodes in the network including itself.

The results of the internal health monitoring are made available locally and via the FlexRay network. As a result, the cross-checking among nodes can detect faulty components. Still, the health monitoring systems implemented on the network nodes of MOBILE only serve as demonstration examples of such systems.

Content monitoring in the demonstration application covers the plausibility checking of value ranges and the analysis of the proper forwarding of the driver demands to the actuators. The first component is realized by simple and reliable threshold-check algorithms. The second part can be implemented in different ways: a parallel command derivation from the input data on both the primary and the secondary node allows to directly compare results. This approach is expensive as it requires redundant wiring of the controllers to all sensors and actuators and diverse implementation of the same software. In exchange, a high level of precision is achieved. But in MOBILE, the black box structure hinders the application of this approach. As a result, the secondary node only monitors the safety-relevant in- and outputs of the primary node. It relies on the knowledge about the surrounding vehicle rather than about the algorithms executed by the primary node to diagnose the system. The key factor that is considered in the demonstration application

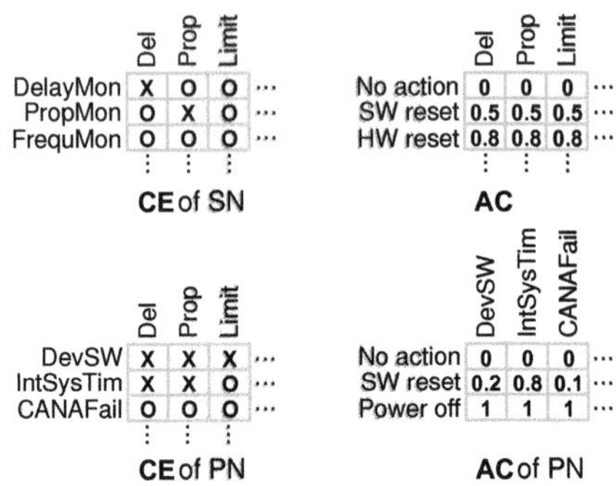

Signal groups to indicate:
Del: Delay in processing of signals detected; Prop: Proportional deviation in signals detected; Limit: Signals exceed acceptable bounds;
Faulty components:
DelayMon: Failure of delay monitoring algorithm; PropMon: Failure of monitoring algorithm to detect delays; FrequMon: Failure of frequency monitoring algorithm; DevSW: Failure of software under development; IntSysTim: Failure of interrupt or timing system;
Actions:
No action: no action; Power off: power off; SW/ HW reset: soft-/hardware reset

Figure 7.5.: Excerpts of the example cause-symptom correlation (**CS**) and action-cause correlation (**AC**) matrices of the primary (PN) and secondary (SN) node

is the proper forwarding of the user commands. In order to monitor the proper command forwarding, the secondary node measures delays and proportional errors between the signal curves, which are acquired at the outputs and the inputs of the primary node. The bounds of acceptance for the monitoring are set wide so that the applications on the primary node are little restricted. Again 2 to 10 of these inputs are combined into each signal group.

Matrices

For these input signals, the cause-symptom correlation matrices (**CS**) and action-cause correlation matrices (**AC**) are generated. Figure 7.5 provides an example excerpt of the used matrices. The appendix (Sec. A.4) lists the full matrices implemented for the demonstration example. The elements of the matrices are determined based on the long-term experience with the hardware platform and the software modules for data exchange. The conditional probabilities to determine which of the two nodes in a fault tolerant unit caused a symptom are set specifically for each signal group. In general, the primary node is the "favored" source of error due to higher code complexity and less software testing. To react to faults, PFDH can remain silent ("no action"), command a reset, or power-off the secondary or primary node via a request to the power control units. A software reset restarts the timing system of the operating system. A hardware reset is comparable to power-cycling and is only valid for the secondary node, because the primary node may execute unintended initialization routines. Obviously, the list of actions can easily be extended. In particular, actions that are dedicated to a specific fault are desirable.

Verifying the Inputs of PFDH

To test PFDH, the outlined algorithms were implemented in "C" and executed cyclically at a frequency of 250 Hz. The following paragraphs briefly outline the test results for the used symptom generation algorithms, and then focus on analyzing dedicated decisions of PFDH to check for plausibility and compliance with the given requirements.

The Determinism monitoring: The proper operation of the frequency monitoring algorithms was confirmed on the test bench for the components of MOBILE by generating signals with various transmission frequencies on the primary node and logging the error flags generated by the monitoring unit. All signals with frequencies outside predefined bounds were detected. The applied monitoring algorithms used thresholds to detect out-of-bound conditions. More details can be found in the thesis of Balkan [2011]. Proper operation of the online self-diagnostics of the network nodes was validated by structured fault injection into the code, such as outlined by Madeira et al. [2000]. Additionally, long-term tests with MAX, the test bench, and some tests carried out with Gerdes's X1 vehicle [Bergmiller et al., 2011a; DDL, 2012] contributed to verify the proper operation of the diagnostic algorithms. The test scenarios focused on the detection of overload, scheduling errors, hardware disturbances (e.g., bad/intermittent contact), peripheral errors, and on the disconnection of network nodes. Details on the onboard monitoring can be found in Bergmiller [2008].

The content monitoring: The algorithms for proportional and delay-free forwarding of user commands were tested in simulation with user input data from real test runs on the test bench. The primary node was used to deliberately manipulate the data before forwarding the commands to the actuators. Random influences and regular error patterns were examined. Thereby, again simple threshold monitoring algorithms for timing and proportional deviation were used Balkan [2011]. More complex content monitoring approaches have not been covered in this thesis.

Plausibility of the Decision Making of PFDH

After the proper operation of the monitoring algorithms, which provide the inputs for PFDH has been tested based on the test bench, the plausibility of the decision making of PFDH is investigated in simulation. Therefore, the decision making is discussed for some example scenarios, and consequently a statistical analysis of the decisions of PFDH based on a large number of symptom patterns is provided. The main goal of these tests is to visualize that the system operates as intended and fulfills the set requirements. It is also important to check how the system reacts to situations that were not explicitly considered by the developer during the

	SN responsible: 10%								SN responsible: 33%			
Case	*A*		*B*		*C*		*D*		*E*		*F*	
p_{del}	33%		100%		33%		0%		100%		0%	
p_{prop}	0%		0%		33%		33%		0%		66%	
Node	SN	PN	SN	PN	SN	PN	SN	PN	SN	PN	SN	PN
t_1	NA	NA	NA	**OFF**	NA	**SWR**	NA	NA	NA	**SWR**	NA	NA
t_2	NA	NA	-	-	NA	**OFF**	NA	NA	OFF	**OFF**	NA	**SWR**
t_3	NA	**SWR**	-	-	-	-	NA	**SWR**	-	-	**HWR**	OFF
t_4	NA	**OFF**	-	-	-	-	NA	**OFF**	-	-	**OFF**	OFF

NA: no action; **SWR**: software reset; **HWR**: hardware reset; **OFF**: power off;
p_{del}/p_{prop}: symptom probabilities for signal groups monitoring delay and proportionality
SN: secondary node; **PN**: primary node

Table 7.2.: Action sequences for persistent faults

configuration of the system. To start with, Tab. 7.2 provides an overview of some action sequences that were derived by PFDH when permanent faults were set at the inputs. These sequences confirm important features of PFDH:

- In case of distinct error conditions, the appropriate action is derived quickly (cases *B*, *C*, and *E*) → R3.P2.
- Symptom patterns hinting at a common cause lead to a faster action derivation (case *C* vs. cases *A*, *D*) → R3.P2.
- In case of uncertain error conditions, e. g, due to sporadically occurring single errors, PFDH does not overreact but awaits a confirmation (cases *A* and *D*) → R3.P3.
- Actions with little collateral damage that target the detected faults are preferred to more general actions with higher impact on the overall system (cases A, C, D, and E) → R3.P1.
- Conditional probabilities to determine the error perpetrator are important, but the final decisions of PFDH are robust in that regard if meaningful symptom patterns are detected (case *B* vs. case *E*) → R3.P3.

To complement these individually discussed cases, a statistical evaluation of the actions triggered by PFDH is performed based on varying symptom patterns and probabilities that are fed into PFDH. Figure 7.6 illustrates the results of a simulation set. Each subplot indicates the actions commanded by the decision unit if each signal group is associated an error probability of zero or the value given in the plot. This way, the decision making under the varying conditions of uncertainty can be evaluated. For a better understanding of Fig. 7.6, the following example illustrates the underlying data generation process.

Example: If a system configuration would contain only three signal groups with the error probabilities p_1, p_2, and p_3, the subplot "0 OR 0.10" would be

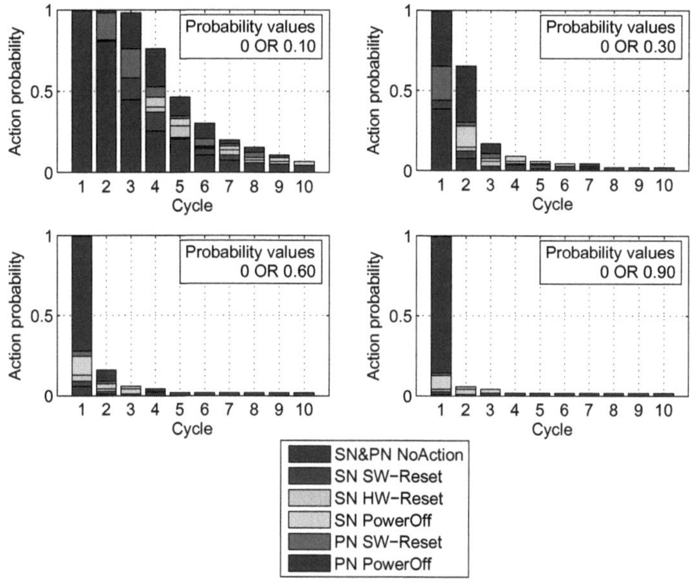

Figure 7.6.: Probabilities of actions applied by PFDH with increasing probabilities associated to symptoms (SN: secondary node, PN: primary node)

generated from the responses of PFDH to the input patterns given in the table below.

p_1	p_2	p_3
0.1	0	0
0	0.1	0
...	...	...
0.1	0.1	0.1

The combination $(0, 0, 0)$ is skipped. The cycle count in the figure indicates the number of consecutive time steps during which the symptom pattern was applied. For the demonstration application, each cycle lasts 4 ms. The bars drawn for each cycle indicate the total probability that there are still two nodes switched on that can be controlled by PFDH. If one node is turned off, the height of the bar decreases. The colored ribbons at each time step indicate the contribution of each individual action to the overall activity of PFDH.

The results of the simulation series confirm what was already outlined for the example scenarios given in Tab. 7.2. High error probabilities lead to a fast action

Discussion of simulation results

derivation while uncertain error conditions are appropriately considered by awaiting a confirmation. The actions with lower impact on the overall system are preferred to more severe interventions if the error probabilities are still "low enough" to tolerate another cycle of delay. Unless at very low error probabilities, final decisions are made within three to six cycles, which is acceptable to assure safe vehicle control at a cycle time of 4 ms (compare Sec. 5.1). Complementary to the results from the example scenarios, the statistical evaluation shows that unclear error patterns, e.g., patterns where a symptom is highly unlikely to be observed on its own without other accompanying symptoms being detected, do not trigger an action even if the failure probability of this individual signal is high. Such symptom patterns are a part of the generic test set, because the patterns were not generated from real world problems but combinatorially. This effect can be seen in the graphs by the small 'no action' bars that are persistent throughout all cycles. An example symptom pattern generating this behavior in the demonstration application would be a single symptom indicating a failure of the FlexRay controller, while the delay and proportional error monitoring algorithms considering the signals transmitted via that data bus do not indicate any errors. Thus, PFDH stays on the "safe side" by performing no action in case of high uncertainty due to implausible input combinations.

Fig. 7.7 is based on the same data set as Fig. 7.6 but highlights a different aspect: it visualizes the probabilities of application of actions within the first and second cycle after a symptom pattern was constantly applied to the inputs of PFDH. The "probability input" value given on the abscissa again references the "0 OR *abscissa value*" scenarios outlined before. It can be seen that with increasing error probability the more powerful actions are favored, while the actions that do not guarantee a solution but have a fair chance to solve a problem and impact the system less, such as a software or hardware reset, are mainly applied at low or medium error probabilities. This indicates that the simple cost function based on the probabilities of success and error and the weighting factor works for the demonstration scenario.

Differentiation of PFDH

Sec. 7.1.1 introduced the main goals of PFDH and outlined some alternative decision making systems with a similar focus. After PFDH has been introduced, the differentiation of PFDH shall be revisited following three main aspects derived from related work:

- The research group of Isermann introduced a diagnostic approach that is applicable to technical systems and clearly distinguishes between causes and symptoms. This supports structured system analysis and suits the separation into faults, errors, and failures introduced in ISO 26262. The matrix based form supports configuration of the system. This idea was adapted and extended by additional matrices to relate detected faults to available actions.

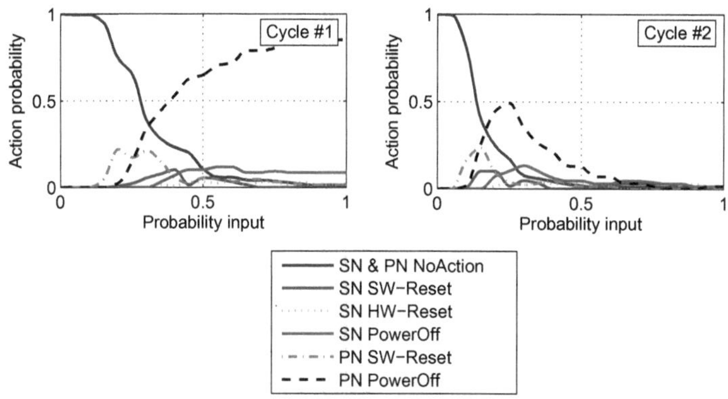

Figure 7.7.: Probabilities of actions for action sequences generated by PFDH (SN: secondary node, PN: primary node)

- Schwall [2005] follows a probabilistic approach using Bayesian Networks. This way, uncertain failure conditions are considered and multiple symptoms are merged to derive possible causes and actions from the resulting probabilities and costs set by the developer. PFDH differs, because several simplifying assumptions are made for computational reasons, which in the first step makes PFDH less powerful. The main added value by PFDH is the comparative approach, which considers both the primary and secondary node and includes multiple actions including "no action". This way, decision thresholds can be generated easier and the failure rates of the diagnostic unit itself are explicitly taken into account.
- Finally, machine learning approaches were touched due to their capability of learning from recorded data. So far, this aspect has not been considered by PFDH. Therefore, and to provide a benchmark with Bayesian Networks, the following paragraphs compare PFDH with a Bayesian Network in more detail.

PFDH and Bayesian Networks

To start the benchmark, a Bayesian Network was set up that performs the cause-symptom correlation and the action derivation in the same way as PFDH. The timely effects are excluded and dealt with externally. Thus, just a "static" Bayesian Network was designed, which suffices to investigate the action derivation process being the core part of PFDH. Fig. 7.8 outlines the structure of the generated directed acyclic Bayesian Network. Starting from signal group error probabilities via the association to the primary and secondary node and the cause identification, the Bayesian Network derives the probabilities of success for actions and determines the associated costs. For the demonstration application, this network features 147 nodes. The equivalence of PFDH and the Bayesian Network can be mathemat-

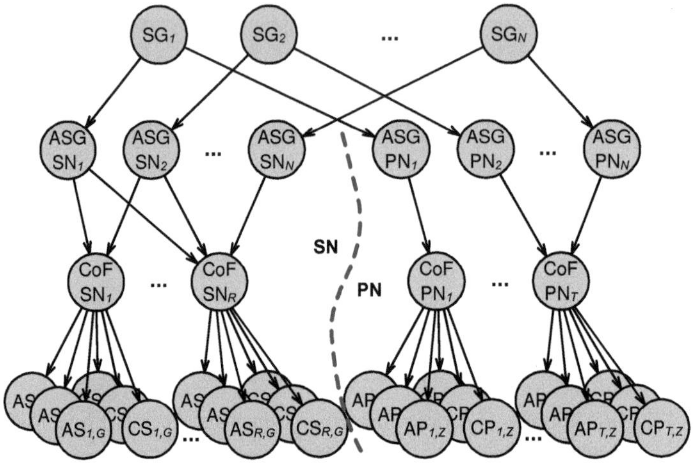

$AS_{r,g}$ / $AP_{t,z}$:	efficiency of action *g/z* to address *r/t* -th cause on the secondary/primary node
$CS_{r,g}$ / $CP_{t,z}$:	costs generated by action *g/z* on cause *r/t* on the secondary/primary node
CoF $SN/PN_{r,t}$:	*r/t*-th cause of failure on the monitoring/development node
ASG $SN/PN_{r,t}$:	signal group failure rate associated to the secondary/primary node
SG_n:	probability of failure detected by signal group n

Figure 7.8.: Structure of the PFDH-equivalent Bayesian Network

ically proven by induction if the conditional probability tables for a node with a given number of parents are structured appropriately (see proof in Sec. A.5). During testing, both PFDH and the Bayesian Network were generating the same decisions, as theoretically proven. Computationally, PFDH makes use of important simplifications derived from the special field of application and thus is able to calculate the results faster if compared to the standard solving approach for Bayesian Networks. The efficiency increase is indicated by the needed number of calculations (see Sec. A.6). As a result, PFDH should need less than 25% of the computational power. During real testing, the Bayesian Network consumed 97% of the calculation time, PFDH the remaining 3%. This difference can just give a rough idea of the computational effort, because PFDH is implemented efficiently, whereas the implementation of the junction tree engine used to solve the Bayesian Network was not examined in detail. Additionally, not necessarily an algorithm for exact inference, such as the junction tree approach, needs to be used to solve the network. For the application, faster approximative solutions may also be sufficient [Murphy, 2001].

The equivalence between the Bayesian Network and PFDH makes it possible to exploit the additional opportunities provided by Bayesian Networks. In particular, learning algorithms developed for Bayesian Networks could be reused to adjust the entries of the introduced matrices based on the recorded data from a fleet of vehicles similar to the approach by Müller [2011]. Then, the entries of the matrix that are pre-defined by a developer could be automatically fine tuned by measurements from a fleet of vehicles and thus improve the quality of decisions made by PFDH. Online learning in the vehicle for improvement of PFDH seems unreasonable not only due to the limitations in computational power, but also because faults are expected to occur rarely or never during the lifetime of the vehicle. Thus, training data is missing locally.

7.1.4. Conclusion

The main goal of PFDH is to address a critical part of the safety concept of MOBILE: a solution had to be found to deal with an ECU as a part of a fault tolerant unit that executes black "box software", which can flexibly be exchanged. Therefore, PFDH provides a matrix based approach that supports an easy configuration and the degradation concept of the vehicle. The probabilistic elements ensure the traceability of the action derivation process while handling uncertainty. This is extended by a novel approach to associate costs to individual actions, which is a vital step towards a strategy of stepwise escalation of actions depending on the current fault state of the system. Using signal groups to generate the symptoms increases the robustness of the approach with regard to sporadic faults and implausible error states, and helps to quickly condense huge amounts of available measurements. For implausible symptom patterns, PFDH acts conservatively while deriving actions quickly under clear error conditions, which reflects the behavior of a human expert. As a further main contribution to robust decision making, the unique comparative approach, which considers the actions for both involved controllers including the action "no action", relieves the developer of the rather unintuitive definition of thresholds for intervention. Finally, the computational efficiency of PFDH makes it a viable diagnostic solution for the 1oo2D structure implemented for the fault tolerant units of MOBILE.

Based on the presented simplified demonstration application, the configuration of PFDH can easily be extended to improve the situation specific degradation and the diagnostic performance. In particular, further targeted actions could support the envisioned self-healing concept for the overall system and minimize the disturbance of the remaining system by fault handling. Also, the stability control algorithms that have to be developed for MOBILE could be parametrized by PFDH to support using functional redundancies for functional safety. In summary, PFDH can take over the tasks associated to it by the overall system architecture and thus contributes to the safety concept of MOBILE.

In future work beyond the limits of the MOBILE project, an application of PFDH in series vehicles for monitoring of complex software could be evaluated.

During first discussions with partners from industry, it became obvious that a proof of functional safety for complex components of the electronics system in a vehicle causes huge effort. An accompanying unit for a critical ECU, similar to the "secondary node", could ensure the functional safety and could be reused for different revisions of the monitored software if standardized interfaces or interface description modes are defined. Thus, the functional safety of the secondary unit or co-processor executing PFDH has to be proven only once. Also, the proof may be simpler due to the reduced software complexity. This approach based on a main controller and a companion executing PFDH is similar to the monitoring strategy proposed by the Robert Bosch GmbH for controllers that implement an electronic gas pedal. This approach has been successfully followed for several years [Schäuffele and Zurawka, 2013, p. 212]. Still, the PFDH based approach targets even more complex tasks, where a (complete) diverse duplication of the primary function, as currently done, causes too much effort.

7.2. Cross-Actuator Failure Compensation

The PFDH approach introduced in the previous chapter monitors the vehicle based on various symptoms and derives appropriate actions to handle existing faults. The presented list of actions available to PFDH focuses ways to reconfigure the EE system. So far, an actuator failure has not been considered. To handle such a failure relying on functional redundancies, an appropriate stability control system has to be provided. The stability control can be triggered and parametrized by PFDH actions to improve the intervention strategy for each specific fault. Eventually, the functionality of a failed actuator has to be replicated by intervention of functional "back-up" devices to ensure at least a degraded emergency operation of the vehicle. In over-actuated vehicles, a failure of one actuator to steer a single front-wheel may, e.g., be compensated by coordinated intervention of the remaining steering units and the braking and torque vectoring system. Working on a similar challenge, Ono et al. [2009] summarizes this approach as follows: "If each of the tires can be individually steered and operated for traction/braking, the task of control grows from three control inputs (longitudinal, lateral, yaw) to eight, providing redundancy to the system" [Ono et al., 2009, p. 89]. The degree of redundancy provided by these additional actuators has to be evaluated for all relevant conditions. Especially, when driving at the limits, it will not be possible to compensate all failure modes of all actuators. This is, e.g., indicated by the results of Brown et al. [2007], who show that the effects of differential drive torque used to emulate virtual vehicle behavior is limited due to tire and motor saturation. Ongoing work with MOBILE also shows that some failure modes, such as locking brakes, cannot be compensated by the control system [Stolte et al., 2014]. Still, the appropriate choice of the safe state of a component can help to address this challenge. E.g., open brakes are significantly less critical than locking brakes, or a back-drivable defective steering unit may be controlled by appropriate torque applied to the according wheel to some extent.

Increase benefits, reduce costs

Following this strategy based on functional redundancies, costs could be reduced by economizing redundant components at least for some failures. Additionally, none of the available actuators is limited to stand-by operation, but customer benefits can be generated from the functional extensions provided by each actuator. Still, it must not be forgotten that the identification of the appropriate control strategies to compensate various possible actuator failures under uncertain environmental conditions is challenging and scientifically not fully solved. As pointed out for some failure-scenario combinations this might even be impossible. Furthermore, the performance of the control system to handle given failure modes has to be evaluated quantitatively to facilitate the contribution to a proof of functional safety, which has so far not been addressed in research. This section summarizes the first steps towards such control systems taken in the MOBILE project and references selected related research projects.

Model-following controller

To start with, the potential of an over-actuated vehicle was examined by designing a controller to make the real experimental vehicle follow a virtual reference. The reference can be modified to evaluate the different handling characteristics simulated by the virtual vehicle during real test drives. Similar works with some restrictions have already been carried out by Asano et al. [1991], Cornelsen et al. [2011], or Laumanns [2007]. Asano et al. [1991] implemented a controller for lateral dynamics with intervention into the rear-wheel steering system to follow a yaw-rate reference given by a virtual model. Laumanns [2007] also focuses yaw-rate control but additionally includes an actuator for front-wheel steering and an active roll-control system, while Cornelsen et al. [2011] propose a system that considers the longitudinal dynamics of a virtual drive train.

Controller onboard MAX

Extending these approaches, a simple controller to both perform longitudinal and lateral control of the 1:5 scaled vehicle MAX was developed in this work. Figure 4.5 (in Sec. 4.2) refers to this controller as driving performance system and outlines the parallelized controller structure to control acceleration, speed, side slip angle, and yaw rate according to a given reference. It was shown that a wide range of virtual models can be emulated to some extent by the over-actuated scale vehicle. Yaw rate objectives were achieved by controlling the difference of the steering angle at the front and rear axle, while side slip angle objectives were targeted by equal parts in the steering angles. This control allocation achieved significantly better performance in scenarios close to tire saturation than an approach performing yaw control purely with the front axle and side slip control at the rear axle, such as the one followed by Laumanns [2007] and also implemented in the MOBILE project for evaluation. For Max, tires tend to saturate frequently due to the low friction surface that MAX is operated on. In that case the later approach followed one control objective very well, but performed poor in following the other objective controlled by the saturating axle. The first approach allows the user to easily set the priority for each control objective, in these scenarios. The dynamic prioritization strategy for lateral dynamics is based on a virtual load of the relevant actuation units. The load refers to the percentage of the adjusting range that would be required to achieve a control objective. Consequently, this percentage may exceed 100% and can be used to compare the achievement of different control objectives. For longitudinal dynamics, the accelerations were in general prioritized. Only during quasi stationary driving at low accelerations, the control accepts small acceleration errors to re-match the reference speed. This strategy was chosen based on the experiences made in the "InDrive" project [Cornelsen et al., 2011]. There, drivers were much more sensitive to accelerations than to speed deviations. Figures 7.9 and 7.10 provide measurements that were taken during an example drive using the simple model-following controller on MAX. During the drive, PID-controllers with speed dependent parameters performed the individual sub-control tasks and an emphasis was put on the side slip angle emulation. The slow following behavior in the yaw rate tracking results mainly from the slow steering system of MAX towards higher steering angles and the latencies in the command transition to the steering motors. Additionally, the required yaw rates and yaw rate changes tend

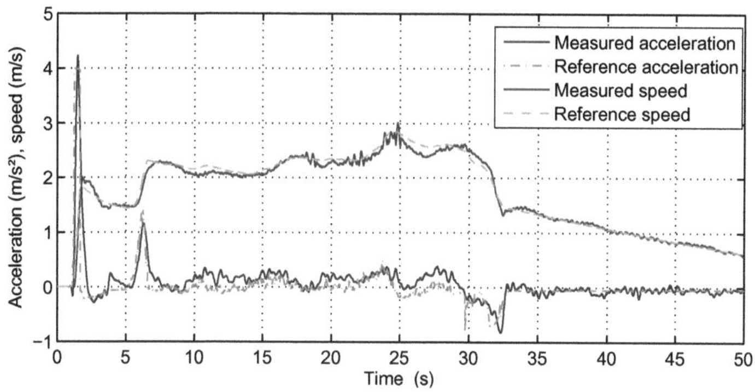

Figure 7.9.: Acceleration and speed of MAX while following a reference

to be higher than in a real vehicle when the scale vehicle is remote controlled. This is due to the stick control of the vehicle and the missing direct feedback to the driver. For further improvement of the control system appropriate feed forward parts or angle dependent parametrization to handle the varying latencies of the steering actuators in the scale vehicle and their limited steering speeds towards higher steering angles could be added[4]. But, especially for this application, MAX is currently only used to demonstrate the basic operation of a controller set-up before migration of the control system onto MOBILE. As a result, the control system was so far not optimized. In summary, the presented controller represents a first step to trigger the implementation of such a system on MOBILE, which features significantly better actuators in terms of power and latency.

Failure compensation

Starting from the promising results obtained with the model-following controller, investigation of suitable compensation strategies for dedicated actuator failures has been started based on MAX. These works are still preliminary and will be an important part of the future research in the MOBILE project. In other research projects, different coordination strategies have been developed so far to achieve various optimization goals. Javadian et al. [2011] and Mokhiamar and Abe [2005] investigate systems that implement yaw rate control by integrated control of the available actuators in an over-actuated vehicle. Still, barely any approaches are found that investigate effects of individual actuator failures on the vehicle dynamics and derive the best suitable control strategy for each scenario. Krüger et al. [2010] introduce BMW's approach to an integrated control system that dynamically allocates the control tasks to individual controllers, which rely on the different available actuators. The system continues to operate in degraded mode if a sin-

[4]The limitations of the MAX scale vehicle with regard to actuation stem from the size and the weight of the vehicle due to the onboard computers and the robust mechanical design. MAX is already equipped with the most powerful steering actuators for remote controlled vehicles that are available on the market.

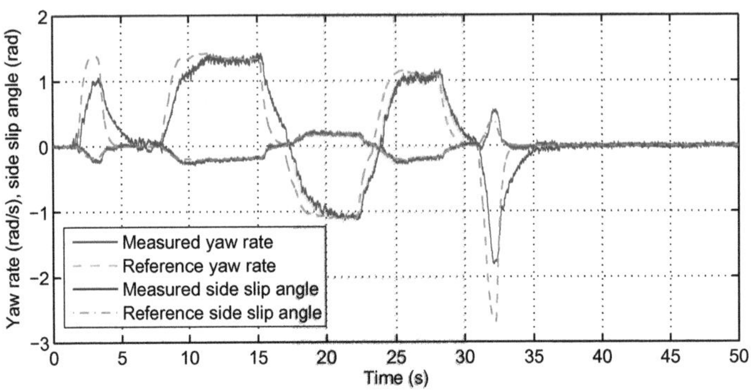

Figure 7.10.: Yaw rate and side slip angle of MAX while following a reference

gle controller fails. But, it does not target the actuator failure itself that may negatively impact vehicle handling. Hoedt and Konigorski [2011, 2013] propose a system for coordinated control of an over-actuated vehicle and analytically derive possible strategies for failure compensation from the modeled degrees of freedom due to over-actuation. Still, the resulting control system is only evaluated in simulation and the failure compensation strategies have not been verified in a real vehicle under varying environmental conditions. Thus, the practical applicability remains to be shown.

Conclusion

In summary, a control system for vehicle dynamics relying on an over-actuated vehicle can significantly influence vehicle handling. Multiple approaches are continuously being developed in research and industry to maximize the benefits from these systems. The research so far has frequently concentrated on individual aspects, certain driving situations, or specific control objectives. Few projects have started to consider the potential of control systems for failure compensation in an over-actuated vehicle. For this, both theoretical and practical experiments are required to identify the optimal control strategy for a given failure scenario including environmental conditions. If appropriate strategies and controllers have been developed, quantifying the success rates of such a control system will become essential. Unlike existing control systems, such as the stability control systems available in modern series vehicles, the control system for failure compensation cannot be designed fail-silent but "always" needs to be available in case of an actuator failure if other redundancy measures are economized. This need for guaranteed operation under varying environmental and driving conditions will put up a significant challenge for future work. Nevertheless, the approaches that will have to be developed to quantify success rates of these algorithms will not only be useful for evaluating functional safety of stability control systems but can also be transferred to the field of automated driving. In this field, similar challenges arise if the performance of an environmental perception system has to be quantified to evaluate functional safety of automatic operation. For this thesis, the conducted experiments and the brief assessment of developments by other research groups suffice to motivate the integration of such a control system into the functional safety evaluation of MOBILE, which then demonstrates the potential of such an approach in terms of the costs, the customer benefits, and the functional safety. Judging from the current developments in stability control algorithms that rely on flexible vehicles, huge further progress can be expected in future work.

8 Strategic Failure Prevention

"In strategy it is important to see distant things as if they were close and to take a distanced view of close things."

Miyamoto Musashi

The successful execution of a complex task requires a high level of skill from the actor. If the quality of the results of the task is additionally subject to external influences, the actor has to react properly to such disturbances to reduce or avoid negative effects. If the task is on top of that real-time critical, such as the driving task, proper reactions have to be derived quickly to ensure successful control. Such quick reactions to unexpected disturbances are challenging, and the faster a complex reaction has to be executed to mitigate negative consequences, the higher the uncertainty of success becomes, in particular, for an untrained actor. If the actor performs the task multiple times, his skill level will improve and thus the probability of success to handle disturbances. Still, with the increasing experience, the actor will also increasingly identify the most important influences on his task, their type, effect, and how to deal with them. Some of these influences will occur sporadically and unforeseeable for the actor, but some will occur on a regular basis or triggered by a certain event or subtask. Instead of performing the task ever the same, as he did the first time, the actor will naturally improve his "plan" how to perform the task and determine which situations to avoid. Usually, it is more comfortable to avoid a failure than to handle the consequences. If this planning achieves a high level of abstraction and the actor decides "a-priori" what aspects have to be taken special care of and how this has to be done, this work would refer to the resulting concept as the actor's strategy.

Strategic mechanisms for MOBILE

Starting from the sketched example, this section introduces approaches to strategically prevent failures in highly flexible vehicles as MOBILE. These strategic mechanisms reduce the probability of occurrence of critical situations that have to be dealt with based on the tactical mechanisms introduced in Cha. 7. Strategic measures will require to consider long-term effects on the vehicle and to abstract the driving task and the basic properties and skills of the vehicle-driver system (the "actor"). When this is done properly, the resulting systems can reduce the probability of failure of the vehicle, protect the passengers and the environment, and

P.J. Bergmiller, *Towards Functional Safety in Drive-by-Wire Vehicles*,
DOI 10.1007/978-3-319-17485-3_8

also increase the efficiency and convenience. Of course, these strategic mechanisms cannot reduce the probability of occurrence of risky situations down to a level that would comply with the functional safety standards if tactic measures are skipped. But, the tactical systems are unloaded by reducing the number of critical situations that are unlikely to be treated perfectly. In any case, the comfort, the reliability, and the driver confidence in the vehicle are increased if, e.g., situations that necessitate a transition of the system into emergency run modes are avoided.

This chapter introduces two strategic mechanisms for application in vehicles: a system for long-term online optimization of control and a structured approach to self-representation[1] for the vehicle. The online optimization system targets long-term wear and load balancing among different types of components to reduce the probability of failure of critical system components. The self-representation adds an additional degree of abstraction to the description of the vehicle by providing a hierarchical view on properties and skills of the ego vehicle, which can then be used by a decision making system or the driver for the strategic planning of tasks.

8.1. Online Optimization for Load And Wear Balancing[2]

Modern automobiles feature multiple actuators and power consuming units. Proper coordination of these units is important to optimize the energy efficiency and to avoid the overloading of individual components, which might eventually cause failures. As a result, the lifetime of the vehicle can be increased by an appropriate coordination unit. In full electric vehicles, the situation becomes even more challenging: on the one side, electric vehicles feature major drawbacks compared to gasoline engines. E, g., the current battery technologies limit the energy available for the driving operation and are sensitive to misuse. On the other side, a full electric drive train can provide additional degrees of freedom for the vehicle control, e.g., the option to drive wheels individually or to recuperate the braking energy [Yu et al., 2009; Ringdorfer and Horn, 2011]. If MOBILE is considered, the four-wheel steering and drive, the electric braking system, and the State of Health/Charge of the independent drive batteries need to be considered. Protecting these components contributes to the reliability and functional safety of the vehicle.

Limitations of the current approaches

As will be outlined, the current optimization and balancing approaches in series vehicles and research projects hardly target the balancing of wear and load among different types of components with regard to clearly defined optimization goals. An important reason why current systems only treat individual aspects and neglect the vehicle-wide interconnections is most likely the exponentially increasing complexity of the decision making system if the interdependencies between the traditionally separated technical domains are considered.

[1] Proper definitions of nomenclature will be given in Sec. 8.2.

[2] Parts of this section have been pre-published by the author in Bergmiller et al. [2012] and Bergmiller et al. [2011c].

Table 8.1.: Requirements for the online optimization system derived from top-level requirements for MOBILE given in Tab. 4.1

R1	**Mechanical and electronic modularity**
R1.O1[a]	The online-optimization system shall be designed modularly to support the flexible integration of new optimization goals into the system.
R3	**Functional safety**
R3.O1	Unintended effects of the optimization system on the vehicle handling must be avoided.
R3.O2	The number of critical failures of the system components shall be reduced by preemptively avoiding overloads and unequal loading.
R3.O3	Failures of the optimization system must not affect the controllability.
R4	**Limited degree of hardware redundancies**
R4.O1	The resource consumption of the system has to be limited and shall be adaptable to the intended application scenario.

[a] R1.O1 stands for the first requirement on the online-optimization system derived from the top-level requirement R1.

To demonstrate the potential of an optimization systems that operates across different domains, a suitable system architecture is presented that allows to address the complexity challenge with simple and readily available off-the-shelf components. Thus, the core contributions of this section are the approach to operate across domains and the system architecture rather than the individual algorithms to model the system, predict changes, and solve the resulting optimization problem. One strength of the chosen approach is that existing solutions from different domains can easily be combined. The operability of the proposed system is verified using an example application that balances load among components of the drive-train, the tires, and the steering system. The development is driven by the requirements given in Tab. 8.1. Most importantly, the flexible optimization system shall reduce the number of critical system failures by considering cross-domain dependencies. Therefore, the system contributes to functional safety but also reduces the effort for maintenance of the vehicle and thus increases the driver comfort. Requirements

To address these challenges, this section proposes an approach to optimal long-term balancing across different domains (e.g., tire wear vs. battery state) in an over-actuated vehicle based on classical optimization algorithms (Fig. 8.1). The optimization system "parametrizes" the locally operating control systems, which makes it possible to meet the real-time requirements and to focus on the "vehicle-level" context. Following this approach, also a slight reduction of the overall wear seems possible. Still, this is not the main focus of this thesis, because appropriate Solution strategy

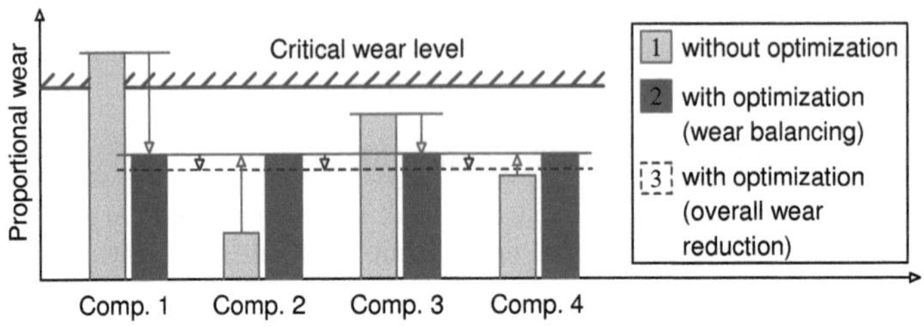

Figure 8.1.: Balancing of wear of multiple components and overall wear reduction

equipment to measure wear precisely for benchmarking is not available. The simulation experiments and the real test runs with the scaled vehicle MAX (Sec. 4.1.2) demonstrate that the approach is applicable in real-time critical scenarios. The example system focuses on balancing of the tire wear while taking the battery charge levels and the actuator temperatures into account. This demonstration application was motivated by the observed unequal wear of the left and right tires when driving with a fixed axle geometry. Amongst others, unequal wear of tires results from the typically higher numbers of turns in one direction than in the other direction during lifetime of the vehicle (more left hand turns than right hand turns for right-hand traffic [Singh et al., 2012]). Additionally, the real-time criticality of this balancing task is high compared to other possible fields of application of the proposed optimization system, which makes the use case a good benchmark scenario.

8.1.1. Related Work and Contributions

Reduction and balancing of wear as well as protection of expensive components is an important goal in the design of modern vehicles. The commonly applied measures target both the mechanical design and the vehicle electronics. The following list highlights some important aspects:

- A proper design of the steering geometry reduces the tire wear while driving curves, e.g., the Ackermann steering geometry is proposed for the driving at low speeds [Isermann, 2006; Mitchell et al., 2006].
- The Battery Management Systems are targeted in multiple research projects and ensure the safe operation of the drive battery in electric or hybrid electric vehicles by balancing the cells of a battery pack or limiting the in- and outgoing currents [Chen et al., 2008; Sen and Kar, 2009; Zheng and Zhao, 2009].
- The energy management units prioritize the power demands of comfort electronics and control the energy distribution in the onboard network [Reif, 2010; Robert Bosch GmbH, 2008].

- Most car manufacturers provide start-stop systems for their cars, which switch off the combustion engine during short halts of the vehicle to save fuel and reduce the emissions if a restart can be guaranteed [Zhong et al., 2010]. Similar control strategies are developed for the usage of the combustion engine in hybrid electric vehicles [Yang et al., 2010].

Need for coordination

This list of actions represents an excerpt of measures taken in modern vehicles to balance the wear, protect the components, reduce the energy consumption, and ensure the operability of the vehicle. As can be seen, all measures target at a well defined field of application. The potential of a vehicle-wide operating coordination system that includes measures from various fields of application is not yet exploited.

In operations research, similar optimization problems have to be solved to maximize the output of industrial plants while dealing with the restricted resources. Apart from real-time requirements, these challenges are comparable to the ones in modern vehicles. For example, Alcaraz and Maroto [2001] and Zhang et al. [2008] discuss different optimization algorithms to minimize the lead times in the production processes with mutual dependencies and parallelized actions while obeying the multiple constraints for the production capacities and sequences. Coello et al. [2007] summarizes several further research projects in different fields of application that rely on optimization algorithms. These include approaches that optimize vehicle layout during design time with regard to fuel consumption and emissions [Coello et al., 2007, pp. 415]. In general, multiple optimization algorithms have been developed to identify optimal solutions. A brief summary of some basic characteristics of these optimization approaches will be given in Sec. 8.1.3 when an appropriate strategy for the optimization task addressed in this thesis will be defined.

Modeling components

Based on a suitable optimization algorithm, the optimization system proposed in this section adds a coordination layer at vehicle level for further improvement of the energy and load management. As a basis for the implementation of the demonstration application, models to estimate the tire wear, the temperatures, and the battery states are needed. For the batteries, simple input/output models based on the consumed power and the charge inputs are used. Similarly, a rough approximation of the temperature profiles by first-order lag elements serves to predict the temperatures of motors and power electronics of MAX. For the tire wear estimation, a more complex approximation model is introduced to capture the most important quasi-static influences on tire-wear to compare the results from different test drives. The model is based on the contributions of several researchers.

Impact factors on tire wear

Tire wear depends on several important factors. Li et al. [2011] consider the tire pressure, ambient temperature, speed, sprung and unsprung mass, suspension stiffness and damping, side slip angle, tread stiffness and damping, sidewall stiffness and damping, and road roughness. Taking these influences into account and by modeling the tire with a brush model[3], Li et al. [2011] calculate the lost mass of a

[3]The tire brush model assumes that the tire contact consists of a set of bristles that can be modeled as a series of independent springs that undergo deformation. More details can, e.g., be found in [Rajamani, 2006, pp. 102].

tire based on the frictional power as introduced by Lupker et al. [2004]:

$$\Delta m = f_1 \cdot W^{f_2}. \tag{8.1}$$

f_1 and f_2 represent empirical factors to take the temperatures and the abrasive character of the surface into account. From the lost mass, the reduction in tread depth is calculated based on geometrical relations. Using the resulting numerical model, Li et al. [2011] perform a sensitivity analysis to determine the most important factors for tire wear. According to this analysis, the top three factors are the side slip angle at the wheel, the wheel speed, and the sprung mass of the vehicle. At least these three factors should be covered in a simplified model. All other factors, such as the temperature and pressure levels, have less than half the impact of the least important of the top three. A weakness of the formula introduced by Li et al. [2011] is the missing explicit coverage of the longitudinal slip at the tire. To a certain extent, the longitudinal slip is covered by the speed influence, but especially sharp accelerations are not sufficiently considered.

Learning from measurements

In the project TROWS (2000 to 2005) funded by the European Community, Lupker et al. [2002, 2004] analyze different types of truck tires and tires of regular cars both in the simulation and during real test runs. Lupker et al. [2002] describe the measurements of tire wear that were taken on a test track or on test benches. The measurements confirm that tire wear "non-linearly depends on numerous parameters" [Lupker et al., 2002, p. 1] and is proportional to the contact forces and thus the frictional power dissipated in the contact area. Starting from this assumption, Lupker et al. [2004] build up both a numerical finite elements tire model and a physically motivated simplified model. The finite elements model serves to derive several parameters for the simplified model. In the simplified model, the dissipated energy as input to Equ. 8.1 is calculated based on longitudinal and lateral slip [Lupker et al., 2004, p. 172]:

$$W = F_L \cdot (\Omega \cdot R \cdot \cos(\alpha) - \omega \cdot r) + F_T \cdot \Omega \cdot R \sin(\alpha). \tag{8.2}$$

In the formula, F_L and F_T are the measured longitudinal and lateral contact forces, $\Omega \cdot R$ the speed of the abrasive disc at a test bench, which simulates the speed of the vehicle, α the side slip angle, and $\omega \cdot r$ refers to the wheel speed. Based on measurements, Lupker et al. [2004] additionally derive that it suffices to sample the wear once a second to capture the quasi-static long-term wear of a tire.

Modeling camber

Also starting from a tire brush model, Huang et al. [2010] derive a physically motivated tire wear model, which models the contact patch as a trapezoid with varying dimensions depending on camber and normal forces. This patch is split into slices that are oriented in longitudinal tire direction (Fig. 8.2). For each slice, a pressure distribution and the longitudinal and lateral forces are derived. The tire wear for each slice is again estimated based on the dissipated power calculated from the forces and speeds in longitudinal and lateral directions. To fit the model to real tires, empirical data is added, e.g., to configure the longitudinal and lateral stiffness of each tire element.

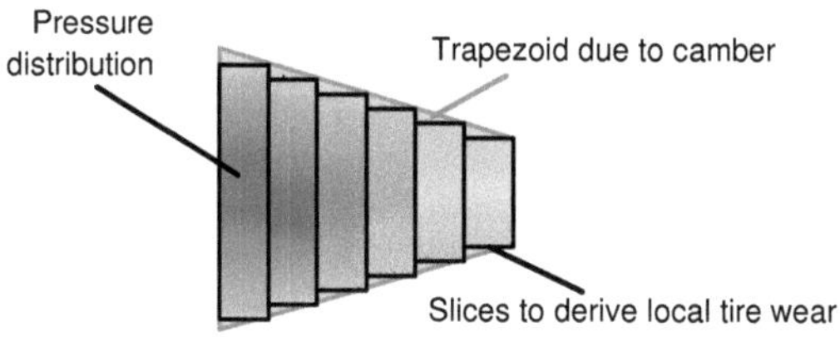

Figure 8.2.: Trapezoidal contact patch split into slices for calculation of local tire forces according to Huang et al. [2010]

Also starting from a FE model [Gruber et al., 2012a], Gruber et al. [2012b] derive a detailed physical tire model. Several core influences are covered by the model: different geometries of the contact patch as already introduced in the previous paragraph, non-isotropic distribution of shear forces, flexible carcasses, and lateral variation of the rolling radius. The resulting model features the highest complexity when compared with the models that have been introduced so far, and performs well when benchmarked with the FE model. Still, it relies on several parameters determined using the FE model or detailed measurements and thus requires a profound knowledge about the modeled tire. In the MOBILE project, a strongly simplified tire model for real-time application is needed, which can be parametrized based on the limited knowledge about the tires of MAX. The model has to cover basic elements that are a part of all the so far outlined approaches. Details of the accordingly adapted model will be given in Sec. 8.1.4.

Non-isotropic shear forces

Based on the models to estimate the tire wear, the motor temperatures, and the States of Charge of the battery pack, the optimization task has to be performed. In the literature, no works were found that focus on the balancing of wear and load across the different domains in the vehicle (tire wear, batteries, temperatures). Nevertheless, the works of Mokhiamar and Abe [2005, 2006] and Ono et al. [2009] are referenced. Both target the optimal usage of tire forces during critical driving scenarios to provide optimized traction. Abe et al. [2013] extend their approach to reduce tire wear, which somehow resembles the approach in this work, but rather focuses short term effects.

Optimizing tire load

Mokhiamar and Abe [2005] derive an optimization problem with eight outputs resembling the longitudinal and lateral forces at the four wheels that can be controlled independently. After adding three constraints for longitudinal and lateral forces and yaw moment, the optimization problem is linearized and solved analytically. Concluding, the eight outputs for the four wheels are transformed into steering angles and drive torques at each wheel relying on an inverse tire model. The experiments within the research project are carried out with a simulator and indicate the improved performance of the overarching optimization system when compared to individually acting vehicle control systems, such as separate control of the front or rear wheel steering. Further research includes a Drive-by-Wire minicar for testing [Abe, 2012; Abe et al., 2013] and demonstrates that the total dissipated energy during a maneuver, and thus the tire wear, can be reduced by the proposed optimization system in real-time. Still, the optimization system does not consider long-term wear, balancing among components, and constraints to generate comparable vehicle handling for optimized and non-optimized driving.

Ono et al. [2009] extend the ideas of Mokhiamar and Abe [2005, 2006] to derive the theoretical limitations of what control performance can be achieved with the force and moment distribution. The main goal is again to reduce the "μ-rate", meaning the degree of saturation of a wheel, and especially unload the tire with the highest μ-rate. Therefore, Ono et al. [2009] rely on SQP[4] and a steepest gradient approach to independently optimize first the steering angles at the wheels and then the loading of the tires. Additionally, the convexity of the optimization problem is shown, and thus global optimality can be guaranteed. Applicability of the presented system was shown during a μ-split braking maneuver with a vehicle featuring a steerable front and rear axle.

8.1.2. System Architecture

Based on the outlined related work to model the important components of the vehicle and to optimize resource usage, an appropriate architecture has to be defined for the intended optimization system. In doing so, it must be considered that the optimization system features the highest safety criticality, because it directly intervenes into vehicle control. These safety aspects are not covered by the so far outlined research projects. To achieve safe operation, the architecture has to be defined appropriately → R3.03. Also, influences of the (permanently operating) optimization system on vehicle handling that might disturb the driver have to be prevented → R3.01. From the developers point of view, a suitable system architecture has to facilitate the easy adaptation of the optimization system to varying tasks → R1.01 and available computational resources → R4.01.

Integrating the optimizer

Figure 8.3 outlines the driver-vehicle system including the components of the optimization system (optimizer and distribution unit). The slanted rectangles indicate information/inputs, whereas the rectangles indicate processing units. The basic operation is as follows: The driver commands are transmitted to a drive controller. The drive controller derives the commands for the actuators. It can perform simple command forwarding to the individual actuators at each wheel or, e. g, emulate the desired vehicle behavior. In the latter case, the vehicle is controlled according to a reference provided by an online evaluated virtual vehicle model as outlined in Sec. 7.2. The commands by the drive controller are passed to the distribution unit. There, the optimizer modifies the commands to balance wear and prevent the overloading of individual components. The modified commands are forwarded to the actuators. Thus, the optimizer does not perform the low-level control but only modifies the inputs of low-level controllers for steering

[4] Sequential Quadratic Programming [Schittkowski, 1985]

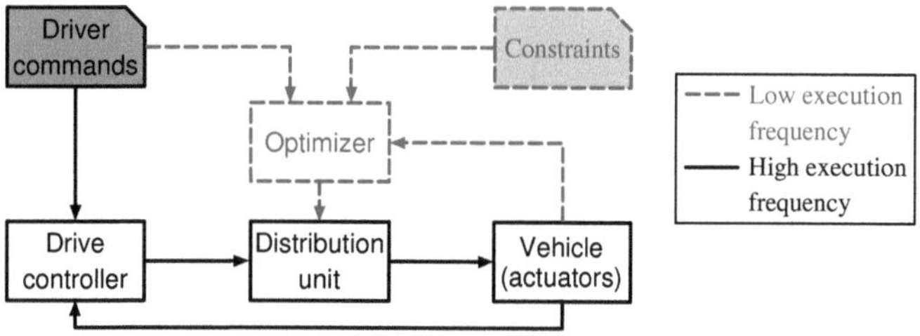

Figure 8.3.: System architecture with optimization functionality

and propulsion. This makes it possible to use the optimization approach both for highly time critical tasks and tasks that rarely require updated outputs. The optimizer is fed with the driver commands, the constraints for the optimization task, and the current vehicle state. Onboard MAX, the optimizer and the distribution unit are executed on a x86 computer and exchange data with the fault tolerant units responsible for actuator control via FlexRay. Only the primary node of a fault tolerant unit regards the transmitted distribution patterns by the optimizer. At any time, the user can switch back to unoptimized driving by either powering off the computer or commanding the primary nodes to ignore the inputs from the optimizer. The proposed architecture provides several advantages that facilitate that the requirements given in Tab. 8.1 are met.

Execution frequency

The execution frequency of the optimizer can be adjusted independently from the execution frequency of the real-time control. If the optimizer does not update the distribution pattern, the distribution unit keeps the previous pattern until an update is received. This is especially useful if the optimization goals change slowly compared to the dynamics of the driver inputs or the vehicle dynamics. Consequently, the computational load can be reduced → R4.01. The experimental section will show that the developer can trade off the quality of the optimization results and the computational load by setting the execution frequency.

The driver

The driver's perception of the interventions by the optimizer is limited for two reasons: the constraints prohibit severe interventions by the optimizer that influence the vehicle handling significantly, and the drive controller can compensate the remaining influences. As indicated by experiments with MAX, selecting the optimization goals properly furthermore contributes to ensure an adequate vehicle handling without noticeable disturbances → R3.01.

Functional safety

The distribution unit monitors the safe operation of the optimizer and can decouple the optimizer from the main signal path for vehicle control if failures occur or during critical driving situations. Therefore, it relies on the already introduced safety mechanisms provided by the remaining network → R3.03. Figure 8.4 details the internal structure of the distribution unit: during the normal operation, the optimizer sets the distribution parameters p_1 to p_Q according to the calculated optimal solution. A monitoring subsystem checks the commanded parameters for

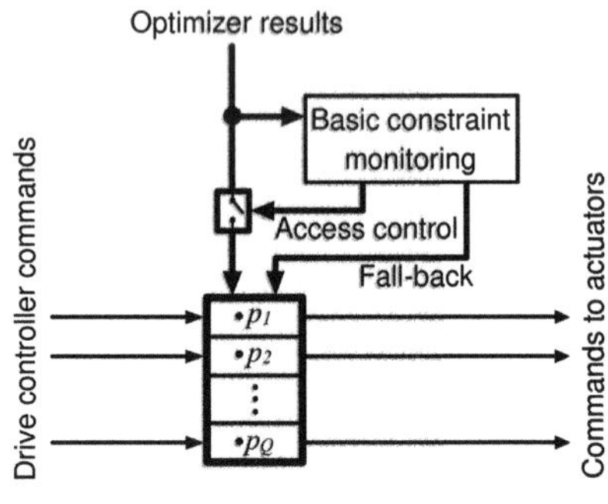

Figure 8.4.: Internal structure of the distribution unit

compliance with basic constraints before they are applied. A basic constraint can, e.g., be that the average of the left and the right steering angle equals the steering angle commanded by the driver. These basic constraints are also included in the constraints provided to the optimizer. To react to faulty patterns, the previous distribution pattern can either be kept or slowly be reverted to the safe default values, which gives the driver or the stability control sufficient time to react to the resulting changes.

8.1.3. Selection of an Optimization Algorithm

As given in Tab. 8.1, the chosen application puts special demands on the optimization algorithm. Consequently, an off-the-shelf algorithm has to be selected that combines a good real-time capability with a high flexibility with regard to adding new optimization goals while approximating a globally optimal solution. To start the evaluation, some relevant features of the optimization algorithms are reviewed.

Single- vs. Multi Object Optimization

In general, optimization algorithms can be subdivided into approaches that focus on optimizing a single objective function and approaches that take into account multiple objective functions [Coello et al., 2007, p. 7]. Single object optimization can be summarized by the following equations:

$$min\ z = f(\vec{x}), \tag{8.3}$$

with the constraints:

$$g_i(\vec{x}) \leq 0; \quad h_j(\vec{x}) = 0; \quad \vec{x}_l \leq \vec{x} \leq \vec{x}_u; \tag{8.4}$$

$$i \epsilon \{1, ..., I\}; \quad j \epsilon \{1, ..., J\} \tag{8.5}$$

and

$$\vec{x} = (x_1, x_2, ..., x_G). \tag{8.6}$$

min z denotes the optimization task expressed as a minimization of the objective function f with the vector of input variables $\vec{x}$ that is modified to identify an optimal solution. Any solution has to obey inequality and equality constraints given by the sets of functions g_i and h_j with I and J elements, respectively. $\vec{x}_u$ and $\vec{x}_l$ mark the upper and lower limits for each component of $\vec{x}$. Multi object optimization extends the optimization objective as follows:

$$min\ z = (f_1(\vec{x}), f_2(\vec{x}), ..., f_N(\vec{x})). \tag{8.7}$$

f_1 to f_N represent the N objective functions considered by the Multi Object Optimization. The definition of constraints remains as introduced.

Evaluation

Both optimization approaches can be applied for the intended application. Single Object Optimization requires all optimization goals to be integrated into one objective function, Multi Object Optimization allows to define one objective function per goal. For complex optimization problems with multiple goals, it is challenging to integrate all factors into one objective function. In particular, a cross-compensation among optimization goals (strongly improving one factor compensates the degradation of another factor) has to be avoided. Multi Object Optimization addresses this problem. Depending on the configuration, it targets at an "overall compromise". No objective is fully ignored while others improve. Also, the effort for adding an additional optimization goal decreases, because the user does not have to reconsider the overall system or deal with the mentioned cross-compensations. But, Multi Object Optimization generates a pareto frontier of optimal solutions, and the user has to provide a selection algorithm [Coello et al., 2007, p. 275]. Defining this selection algorithm can be challenging and significantly impacts the quality of the optimization results. In the following paragraphs, a viable strategy to address this challenge for the given demonstration application will be introduced, and thus Multi Object Optimization was chosen for better usability and extensibility.

Evolutionary vs. Gradient Based Optimization Algorithms

Basically, evolutionary and gradient based optimization algorithms are available to solve the Multi Object Optimization problems. A rough evaluation of both strategies gives reason for the selection of a gradient based optimization algorithm for the introduced application. As will be explained, the gradient based approach is better suited to trade off the real-time and performance requirements of the application.

Evolutionary algorithms

The evolutionary algorithms are inspired by nature. Starting from a random population of possible solutions, they try to identify an optimal solution by recombining the best solutions that have been found so far. To determine whether a solution is good or not, the fitness of each solution is calculated. The fitness considers, e.g., the conformance to constraints and the dominance[5] of a solution with regard to achievement of the individual goals. During recombination, additional random influences can be added to reduce the chance of getting stuck in a local instead of a global optimum. This way, evolutionary approaches can achieve good performance and are able to identify the global optima even for non-convex problems. A general downside of evolutionary algorithms is the significant computational effort due to the huge number of individuals in a population that have to be stored and evaluated within each optimization step. More detailed descriptions of a variety of available algorithms and derivatives adapted to different types of problem can be found in Bäck et al. [1997], Coello et al. [2007] and Domschke [2005] but, e.g., also in Meyer [2003], Deb et al. [2002] or Tusar and Filipie [2007] for important representatives and adaptations.

NSGA-II

To benchmark the optimization system presented in this work in terms of the identification of globally optimal solutions, the well-known and frequently applied evolutionary algorithm NSGA[6]-II was used. The algorithm can be assumed to almost always find the true optimal solution for the example application, and thus the simulation results that are generated with NSGA-II will help to evaluate the performance of the simpler approaches implemented in this thesis with regard to the quality of the optimization. The algorithm is not real-time capable and thus will not be executed on the real vehicles. An implementation of the algorithm in the version introduced by Deb et al. [2002] is available from the Mathworks "Global Optimization Toolbox" [Mathworks, 2013]. Although NSGA-II is a genetic algorithm[7], the modified version implemented by the Matlab Toolbox is capable to operate directly on the decision variables ($\vec{x}$, "phenotype") and does not need to encode the problem in a series of strings comparable to the chromosomes of humans ("genotype"). This option is exploited by this work, because the convergence of the solution due to independence from the chosen coordinate system can be improved by operating on the phenotypes [Meyer, 2003, p. 32]. In particular, the "crowded-comparison operator" increases the efficiency of NSGA-II compared

[5] In terms of dominance, strict dominance, dominance, and weak dominance are distinguished [Coello et al., 2007, p. 244]. A solution dominates another solution if it is better than the other solution in all goals (strictly dominates), or at least one goal (dominates) while not being worse than the other solution in any goal. If a solution is just "not worse" than another solution in all optimization goals, it weakly dominates the other solution. If neither of the given categories applies, the solutions are incomparable or indifferent if all values with regard to the objective are identical.

[6] Nondominated Sorting Genetic Algorithm [Deb et al., 2002, p.1]

[7] A genetic algorithm can be understood best, when compared to genetics in biology. Basically every solution to an optimization problem is represented by a "chromosome", which encodes the solution, e.g., as a bit string. Now the chromosomes from a population that generate the best solutions are chosen by the algorithm. Then, typically two "parent" chromosomes that are selected from the best solutions are recombined according to a given strategy (just like in biology) to generate a new chromosome. Additional random changes of individual "bits" facilitate that the optimization process covers the whole solution space. This process is repeated until a new population is available, and the overall algorithm restarts with selection of the best individuals until a certain break condition is reached.

to other approaches. This operator improves the performance of NSGA-II to generate a uniformly spread-out pareto-optimal front and supports selection of solutions to be kept in the population that have equal dominance levels if too many solutions have been found so far. In contrast to a classical approach that uses a sharing parameter to share fitness among closely neighboring solutions (see, e.g., Hiroyasu et al. [1999]), the crowded-comparison operator improves computational performance and eliminates the need to manually define the sharing distance.

Gradient based algorithms

The gradient based algorithms aim at identification of the optimal solution to a problem by calculating the derivative of the objective functions. Based on the derivatives, the solution is identified by following the gradient towards the optimal result. As multiple goal functions have to be minimized at the same time, a proper formulation of the optimization problem is important to determine which goal functions are targeted to what extent in order to achieve good results. In this work, a Goal Programming and a MiniMax approach were evaluated. For Goal Programming, the optimization problem $min\ z$ is expressed by the following formula based on the objective functions $f_n(\vec{x})$, the goal value for each function $goal_n$ and the weight ("importance") of each goal $GoalWeight_n$ [Mehnen, 2005, p. 91]:

$$min\ z = \sum_{n=1}^{N}(|f_n(\vec{x}) - goal_n| \cdot GoalWeight_n). \tag{8.8}$$

The formulation of the MiniMax approach is [Domschke, 2005, pp. 69]:

$$min\ z = \lim_{p\to\infty}\left(\sum_{n=1}^{N}(f_n(\vec{x}))^p\right)^{1/p}. \tag{8.9}$$

These formulas are evaluated in every optimization step before calculating the derivative, and thus move the solution vector in the resulting direction. Both of the given formulations seem reasonable for the application scenario. The Goal Programming approach adds up the weighted deviations of the individual goals from the reference, similar to Single Object Optimization with multiple objectives. The MiniMax approach focuses unloading of the most loaded component in each step. The MiniMax approach seems in particular useful for balancing of wear in the vehicle and was also followed by Ono et al. [2009]. When compared with an evolutionary solver, it becomes obvious that the gradient based approach will just generate one solution. Thus, the formulation of the optimization problem "substitutes" the selection from the pareto frontier needed for evolutionary algorithms. In both cases, the chosen strategy significantly influences the quality of the results.

SQP

To execute the optimization based on the given formulas, this work relies on the sequential quadratic programming (SQP) algorithm [Powell, 1978]. SQP outperforms most other gradient based algorithms in terms of execution times [Schittkowski, 1985; Ono et al., 2009]. Compared to evolutionary approaches, gradient based optimizers are usually several orders of magnitude faster, which makes them the only viable approach for real-time execution in vehicles. In turn, the purely gradient based approach is more susceptible to the local minima of non-convex functions. The susceptibility of the gradient based approach to local minima will be addressed by a multi-start strategy that triggers multiple optimization processes from different starting points if the quality of the optimization results becomes too low. Evolutionary algorithms are used for offline benchmarks in simulation.

8.1.4. Optimization Criteria and Constraints

Based on the system architecture and the selected optimization algorithm, optimization criteria and constraints focusing on the tire wear, the temperatures of actuators, and the States of Charge (SoC) of the batteries are defined. The criteria cover both the long-term aspects, such as the absolute wear of a tire, and the short term goals, such as the relative balancing among the tires. The following section introduces the accordingly implemented constraints and optimization criteria.

Constraints

Constraints ensure that the interventions of the optimizer are not noticeable for the driver during normal driving, and the actuator limitations are obeyed. The proposed application implements the following constraints:

- *Temperatures:* The temperatures of the two motors and the associated power electronics of MAX are limited to a maximal temperature T_{max}.
- *Maximum change in optimization variables:* The maximal change per time step of each of the distribution parameters p_1 to p_Q and thus of the optimization variables x_g is limited. The limitations avoid rapid changes in vehicle handling that could be noticed by the driver.
- *Actuator capabilities:* For the drive motors and the steering actuators, maximal values for angles and torques are set. Additionally, restrictions due to the actuator set-up of MAX are taken into account. For example, the coupling of the drive torques at the right and left wheel of an axle due to the common drive motor are covered by appropriate constraints (Sec. 4.1.2).
- *Driver command forwarding:* The mean value of the steering angles at the right and the left wheel of an axle that are commanded by the optimizer has to equal the reference set by the driver. This constraint contributes to making the interventions of the optimizer imperceptible for the driver as long as a linear tire model approximates reality and typical road conditions are present. As driver perception is an important aspect, this constraint is discussed in slightly more detail. Compared to an approach where the total lateral force generated at the front axle has to remain equal, the approach via steering angles features some advantages but also limitations. On the one side the approach via the steering angles keeps the uncertainties of the parameter variations due to different friction coefficients on the road out of the system as long as both tires are affected in the same way. Then,

the driver is confronted with the same changing behavior of his car on low friction as he is used to from the unoptimized vehicle. As soon as, e.g., changes in friction affect one tire more than the other, the implicit assumption of the steering angle constraint that the cornering stiffnesses are identical at both tires can cause the vehicle behavior to noticeably deviate from the unoptimized vehicle that the driver is used to. The same may account if the tire characteristics change nonlinearly, e.g., with increased loading or towards higher slip angles. Still, the slow intervention and little modifications of the optimizer should always suffice to prevent safety critical situations. For comfort aspects, the vehicle dynamics controller emulating the behavior of a reference vehicle could compensate remaining influences. In summary, the chosen constraint reduces the sensitivity of the optimizer to changing environmental conditions for most scenarios and thus also limits the risk of unintended actions that might be noticed by the driver. But, this comes at the price of noticeable interventions in some scenarios as outlined. A detailed evaluation of these factors will also require experiments with the full scale vehicle and a driver onboard under changing conditions. As another constraint to ensure proper forwarding of driver commands, the optimizer must never set higher overall torque references than commanded by the driver.

For the application, the constraints mark safe boundaries for vehicle operation. Nevertheless, activating constraints is avoided, because they take away flexibility from the optimizer to smoothly trade off different sets of design parameters. In Sec. 8.1.4 a dynamic-weighting approach will be introduced that supports the early detection of critical states and the smooth adaption of the system based on the optimization criteria and without constraints becoming active.

Optimization Criteria

Based on optimization criteria, the optimizer evaluates the current situation and predicts the effects of a set of commands on objective attainment. As objectives the wear and load of a component are considered.

Reference torque: The optimizer is allowed to reduce the reference torque set by the driver if important/expensive components of the vehicle have to be protected, e.g., from overheating. An optimization goal ensures that the deviation from the commands set by the driver is minimized while taking the current state of the vehicle into account. As a measure, the average of all torque commands for the motors in the vehicle is related to the reference set by the driver. Accordingly, the normalized optimization criterion $n_M(t)$ for the current point in time t results:

$$n_M(t) = \frac{M_{set}(t) - \sum_{g=1}^{g=G} M_g(t)}{M_{set}(t)}, \tag{8.10}$$

where $M_g(t)$ denotes the torque commanded by the optimizer for the motor g, G the total number of drive motors and $M_{set}(t)$ the total torque demanded by the driver. The normalization guarantees the same scaling for all optimization criteria as a basis for the defined weighting of the criteria relative to each other.

Tire wear: To balance tire wear among the wheels, an estimation formula for tire wear has to be provided, which then forms an optimization criterion. As introduced in Sec. 8.1.1, important influence factors on tire wear are the vertical load, the speed, and the slip. For this work, a simplified formula adapted from the formulas of Huang et al. [2010] and Lupker et al. [2004] is applied. The formula does not, split the tire in different slices but considers it as one part. Also, differences in local wear due to inhomogeneous pressure distributions and shear forces in the tire contact patch as analyzed by Gruber et al. [2012b] are neglected. As a result, only the quasi-static wear of a tire is captured by the model, which suffices for basic comparison between the different wheels and the optimization approaches.

1. The normal forces F_{N_i} for each of the $i = 1..4$ wheels are calculated from the static weight distribution and the dynamic loading due to the acceleration of the vehicle. The dynamics of the suspension system are not considered.

2. The longitudinal slip $s_{l,i}$ of each tire is estimated and predicted. As slip estimation is not focused on, an empirical formula identified based on simulation results and real measurements is applied for each wheel.

3. The lateral slip $s_{r,i}$ is modeled as by Burckhardt [1993] for the driven wheel [Burckhardt, 1993, p. 18]:
$$s_{r,i} = |sin(\alpha_i)|. \tag{8.11}$$
α_i denotes the side slip angle of a tire, which is determined based on the current speed, the side slip angle at the center of gravity, and the yaw rate.

4. Because the optimizer is only supposed to operate during normal driving, a linearized function to calculate the effective friction coefficient in longitudinal and lateral direction is used:
$$\mu_{eff_{\{r,l\},i}} = min(C \cdot s_{\{r,l\},i}, \mu_{eff_{\max}}). \tag{8.12}$$
The proportional factor C was derived from practical experiments with MAX on the test ground. To roughly consider the tire saturation during short term overloads, the upper limit $\mu_{eff_{\max}}$ is set for the $\mu_{eff_{\{r,l\},i}}$.

5. Next, the total power P_i transmitted by each wheel is approximated based on the longitudinal and lateral forces, the speeds $v_{l,i}$ and $v_{r,i}$ in longitudinal[8]

[8]Important: This work calculates the longitudinal speed of a wheel based on the speed of the vehicle at the center of gravity, the slip angle, and the yaw rate and does not use the speed from the rotation of the wheel. This approach is valid for small longitudinal slip values, which are targeted by the optimizer. At the same time the approach prevents "spoiling" of the wear measurements in simulation due to "burn-out" scenarios caused by huge drive torques, the simple tire model, and the missing slip controller.

and lateral direction at the i-th wheel, and the resulting powers ($P_{r,i}$ and $P_{l,i}$) as introduced by Lupker et al. [2004]:

$$\begin{aligned} P_{r,i} &= \mu_{\mathit{eff}\,r,i} \cdot F_{N_i} \cdot v_{r,i}, \\ P_{l,i} &= \mu_{\mathit{eff}\,l,i} \cdot F_{N_i} \cdot v_{l,i}, \\ P_i &= P_{r,i} + P_{l,i}. \end{aligned} \tag{8.13}$$

Based on the transmitted power P_i, the tire wear W_i is calculated by integration as introduced in the state-of-the-art[9]. In the MOBILE project, a linear correlation between the transmitted power and the wear is assumed, because only a simplified relative evaluation of tire wear is performed and detailed measurements of the tire wear for benchmarking are missing. For decision making, the optimizer considers the current wear level and the expected wear within a prediction interval.

6. The wear estimates $W_i(t)$ are normalized within each time step:

$$n_{W_i}(t) = 0.1 + 0.9 \cdot \frac{W_i(t) - W_{min}(t)}{W_{max}(t) - W_{min}(t)}. \tag{8.14}$$

$W_{max}(t)$ and $W_{min}(t)$ denote the maximal and minimal wear at the current time step t. The normalized wear for the tire with the lowest wear is set to be 0.1, the tire with the highest wear is set to 1.0.

Motor Temperature: To protect the expensive components of the drive train, a maximal temperature for the motors and the power electronics in the experimental vehicle is defined. The optimizer tries to keep the temperatures of the components below this threshold. Alternatively, an optimal target temperature could be defined, which might be useful for a warm-up of components after the vehicle start. A first-order lag element serves as a simple model of the temperature development of the motors when a certain torque is commanded [Schröder, 2009a, p. 59]. For roughly predicting the future temperatures, this approximation has proven to deliver sufficiently good results and reduces the computational load. In each time step t, the differential equation of the first-order lag element is solved and initialized with the current temperature $T(t)$ of the motor. The predicted temperature $T(t + \Delta t)$, when a constant torque M is commanded for the time interval Δt, is again normalized:

$$n_T(t) = \frac{T(t + \Delta t) - T_{low}}{T_{max} - T_{low}}. \tag{8.15}$$

T_{low} and T_{max} denote the upper and lower bounds of the acceptable motor temperature.

[9]Experiments with MAX showed that the wear due to lateral slip is significantly higher than the one due to longitudinal slip (approx. $13 - 15$ times more impact). The main reasons for this observation are the high camber of the wheels of MAX when steering and the low air pressure of the tires. This empirical correction factor is considered for the experiments that were carried out.

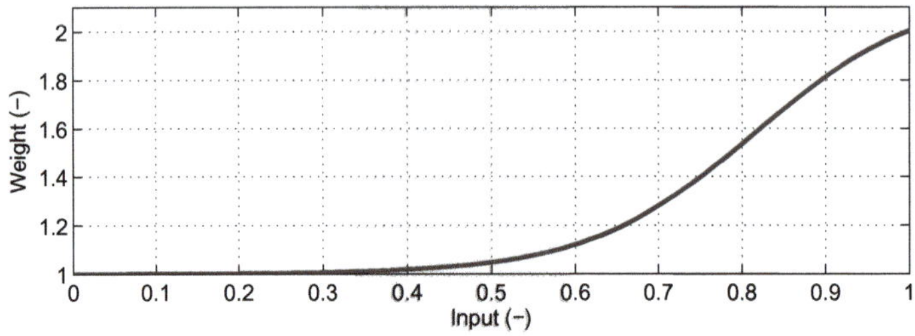

Figure 8.5.: The dynamic weighting factor for optimization criteria

Battery State of Charge: The State of Charge of the batteries is monitored based on the current pack voltage. This is a simplification, but the experimental results have shown that the States of Charge of the batteries of MAX are approximated sufficiently, and the basic operation of the presented algorithm can be demonstrated. More complex battery models to estimate the State of Charge of different types of battery are developed by several research groups, e.g., by Sen and Kar [2009]. The normalized State of Charge n_B of the batteries then results as:

$$n_B(t) = \frac{V_{max} - V_{meas}(t)}{V_{max} - V_{min}}. \tag{8.16}$$

V_{max} and V_{min} denote the maximal and minimal allowed pack voltage, and $V_{meas}(t)$ is the measured voltage. Thus, the normalized value varies between 0 (fully charged battery) and 1 (empty battery).

Dynamic Weighting

To improve the optimization results and to integrate the expert knowledge into the system, the presented normalized optimization criteria are weighted before they are provided to the optimization algorithm. Due to the normalization of all criteria, these weights define the importance of each individual criterion across different types of optimization criteria, e.g., temperature vs. tire wear. The dynamic weights are integrated into the optimization criteria and are thus independent from the formulation of the optimization problem. All weights are calculated according to:

$$dynamic\_weight(\eta) = 1 + \frac{1}{e^{-(K_1 \cdot \frac{\eta}{100} - K_2)} + K_3} \tag{8.17}$$

The parameters K_1, K_2 and K_3 are set individually for each optimization criterion. The input value η represents the percentage of the absolute tire wear, of the maximal acceptable temperature, or of the minimal acceptable pack voltage. Figure 8.5 illustrates the weighting curve for the parameters $K1 = 10$, $K2 = 8$ and $K3 = 0.86$ as used for the given optimization goals. The weighting curve is designed to increase the weight of an optimization criterion if the associated component approaches the absolute wear or energy limit. The flattening of the curve towards the maximal input values limits the weight of one criterion to $1 + \frac{1}{K3}$ even if for some reason one component is used up to a critical level or beyond. The range of output values is chosen to be 1 to 2 within the range of input values from 0 to 1 to ensure that all optimization criteria can be weighted with a predefined "static" factor. Given these constraints for the weighting function, different functions could have been chosen and might work equally well. Basically, this specific function has been chosen as it is smooth, which supports optimization, fulfills the given constraints and rises late but strongly towards the high weight value. The later aspect is important as the weighting should only become active as a criterion approaches a critical state. The static weighting factor multiplies the return value of Equ. 8.17. All components apart from torque deviation are weighted equally (static weighting = 1). Torque deviation is weighted inversely to the temperature, the States of Charge of the batteries, and the tire wear criteria and scaled by a factor of 25 to stress the importance of this optimization goal:

$$static\_weight_{n_M} = 50 - 25 \cdot \max(\{\text{other weights}\}). \tag{8.18}$$

Due to the inverse weighting, torque reduction is alleviated if the state of another component becomes critical.

8.1.5. Experimental Results

To verify the proper operation of the optimizer and to demonstrate that the requirements are met, several test runs were executed both in simulation and with MAX (Sec. 4.1.2). The experiments with MAX, which simulates a standard vehicle with front-wheel steering, demonstrate the real-time capabilities of the system and the basic operation. With MAX, mostly open-loop scenarios were conducted to improve reproducibility. Simulation covers complex scenarios with multiple influences on the vehicle in a closed-loop simulation with a virtual driver. For demonstration, this section references a set of three test tracks:

1. A circle (2m radius) serves to demonstrate the basic operation of the optimizer on MAX at low speeds and for comparison with the Ackermann steering.

2. On a straight line, the proper protection of the drive motors and the batteries is examined.

3. A complex test track with scaled versions of roundabouts and track segments where the vehicle drives at higher speeds concludes the set of tracks. Figures 8.6 and 8.7 show the test track and the resulting speed profile driven by the virtual driver with a simulation model of MAX. The speed limit was set to 5m/s. This track is intended to investigate the reaction of the optimizer to quickly changing conditions and cross effects between optimization goals.

Other tracks and open-loop maneuvers have been examined in the master's thesis of Schuldt [2011].

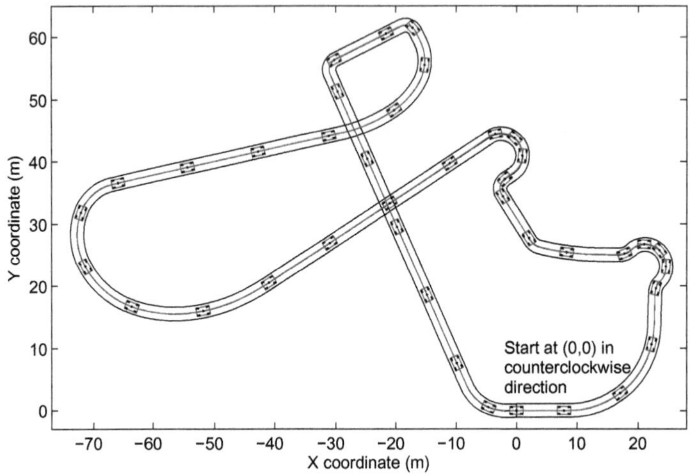

Figure 8.6.: Test track for optimized driving; for better readability, the size of the vehicle is scaled up by two.

The Operation of the Optimizer Onboard MAX

To demonstrate the operability of the optimizer onboard MAX, a simple test run along the mentioned circle was carried out (two consecutive turns with a constant steering angle counter clockwise). While driving, the optimizer was fed with temperature readings, information from the inertial measurement unit (speed, yaw rate, slip angle) and the measurements of the battery voltage levels. Based on these inputs, the optimal steering angles for the front tires and the torque commands for the two drive motors of MAX were derived. It is important to note that for tire wear estimation, the rear axle is considered as one unit, because the optimizer is not provided with control elements to balance the wear between the tires of the axle. For benchmarking, the same drive was conducted using an Ackermann steering and an equal torque distribution between front and rear axle. For the optimizer, a constraint ensures that the mean of the left and the right steering angle at the front axle equals the mean of the Ackermann steering. This generates comparable handling of the experimental vehicle during the test runs with and without optimization. The speed reference for the speed controller was set to 1.1m/s equaling a speed of approx 20km/h of a full-scale vehicle. For such low speeds, the Ackermann steering serves as a good reference for tire wear balancing. During the test run, the temperatures of the drive motors and the States of Charge of the batteries remained uncritical, and thus did not influence the optimization results. The SQP-based optimizer was executed every 0.15s and took approximately 0.1s to calculate on the 1.6GHz Intel Atom D510 platform. To describe the optimization

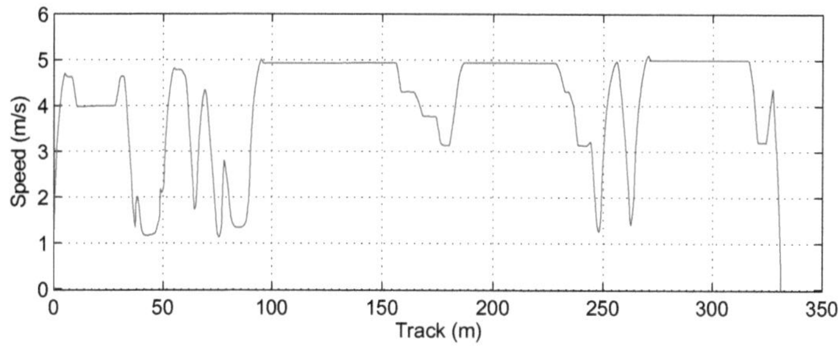

Figure 8.7.: Speed profile generated by the virtual driver for the test track given in Fig. 8.6

problem, both the MiniMax and the Goal Programming approach were evaluated. The evolutionary optimization approach (NSGA2) with a population size of 80 and a maximal number of 100 generations would take approx 36s to calculate the solution required for one cycle (0.15s) and thus is not real-time capable.

Measurements with MAX

Figure 8.8 (top) shows the steering angles at the front tires during the test drive. It can be seen that the optimizer commands a slightly higher angle difference between the tires than the Ackermann steering. This difference results mainly from weight shifts that are not considered in the fixed Ackermann geometry but influence the tire wear. Thus, the vehicle running the optimizer manages to generate a slightly better performance than the Ackermann steering in terms of wear. Tab. 8.2 summarizes the results using some characteristic values: the total tire wear of all tires $\sum_{\forall i} w(i)$ during the drive remains approximately constant for all configurations. Small deviations result mainly from the slight differences in the drive torques during acceleration segments, because the optimizer may reduce the torque by a few percent $(1-5\%)$ to reduce the wear. Of course, the drive controller achieves the same speed, but load peaks during accelerations are reduced. In theory, nonlinearities in tire wear cause an over proportional increase in wear if a single tire is loaded more than others, which is confirmed by studies [Singh et al., 2012]. Thus, the optimizer theoretically can generate wear reduction as also shown by Abe et al. [2013]. Still, the simplified tire wear estimation formula applied in this thesis hardly captures these nonlinearities and thus cannot provide reliable benchmarks on the absolute wear. A significant influence of the optimizer becomes obvious when the standard deviations $\bar{\sigma}$ calculated among the wear of the four tires are analyzed. Both formulations of the optimization task achieve significant reductions. Because the Ackermann steering cannot influence the torque distribution between the front and the rear axle, an additional standard deviation among the wear values at the front axle $\bar{\sigma}_{\text{front}}$ is given for fair comparison, which is also reduced by the optimizer. Finally, the wear of the "worst case" tire ($\hat{W}$, front right tire) is reduced, which is the main contribution of the optimizer in terms of safety and reliability.

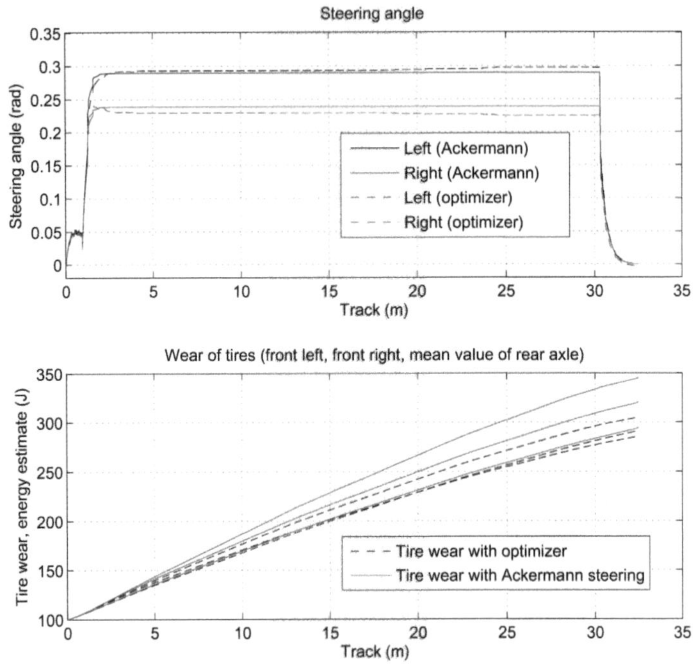

Figure 8.8.: Driving a circle with MAX, MiniMax

Figure 8.8 (bottom) indicates the increasing wear of all tires. The three graphs start at a base wear of 100J and indicate the development at the front left and right tire and the average wear at the rear axle. Obviously, the optimizer keeps all graphs closer together and reduces the wear of the heaviest loaded component, which is in this case the front right tire. When comparing the results of MiniMax and Goal Programming, the importance of this formulation becomes obvious. As can be seen, the MiniMax approach performs better than the Goal Programming approach. This seems reasonable, because the MiniMax approach explicitly targets the worst case goal and thus conceptually supports the idea of unloading a stressed component better. The further experiments with the real vehicle and in simulation showed that the Goal Programming approach achieves better results when the input signals for the optimizer become less noisy. Noise peaks may cause the optimizer to consider the less important goals too much, because all goals are summed, whereas the MiniMax approach robustly targets the most important goal due to its "binary" decision making.

Concluding the experiments with MAX, a 56m long drive following an arbitrary course with speeds up to 4m/s, maximal lateral accelerations of around 3.5m/s^2,

Table 8.2.: Ackermann steering vs. optimizer onboard MAX

Configuration	$\sum_{\forall i} w(i)$	$\bar{\sigma}$	$\bar{\sigma}_{\text{front}}$	$\hat{W}$
Ackermann steering	100%	100%	100%	100%
Optimizer MiniMax	91%	40%	79%	84%
Optimizer Goal Programming	95%	86%	82%	91%
Optimizer MiniMax (long drive)	98%	69%	18%	95%

$\sum_{\forall i} w(i)$: total wear, $\bar{\sigma}$: standard deviation of wear for all tires, $\bar{\sigma}_{\text{front}}$: standard deviation of wear for the front tires, $\hat{W}$: maximal wear of a component.

and mainly left curves was carried out. The results obtained with the optimizer are compared to the results with an Ackermann steering geometry and are given in Tab. 8.2. Tendencies are similar, but with higher speeds and accelerations, differences in balancing in particular at the front axle become more significant. For all experiments with MAX, it has to be considered that (a) the outlined scenarios focus on the tire wear as the battery level and the temperatures remained uncritical during the test drives, and (b) only open loop maneuvers could be carried out due to missing tools for an absolute localization of MAX on the test ground. As a result, slightly different trajectories resulted for the individual test cases due to the actions by the optimizer and the varying conditions of the tires and the test track. To address these aspects and to assess more complex driving scenarios, the simulation model introduced in (Sec. 4.2.1) is parametrized to emulate MAX and serves as a basis for offline simulation experiments.

Taking States of Charge and Temperatures into account

To demonstrate the treatment of overtemperatures and low States of Charge by the optimizer, a simple drive along a 350m straight segment is simulated. The reference speed for MAX is set to 6.5m/s. When permanently driving at this speed, the drive motors are close to overheating, because they are operated around their specified continuous power of approx. 30% of the peak power[10]. For the test scenario, the rear battery is assumed to be weaker than the front battery and is drained 25% faster. The virtual batteries empty quickly and are configured to fully discharge during the test run. The optimizer is based on the MiniMax approach. Figure 8.9 shows the speed and torque profiles with and without optimizer. For the run without optimizer, equal torques are commanded at the front and rear axle, and it can be seen that the motors temporarily exceed their maximal temperature of 70 °C during the acceleration phase (top of Fig 8.10). Above this temperature level, the motors get damaged or may fail. The optimizer avoids critical temperatures by

[10] An improved cooling system for the components could significantly increase this continuous power limit.

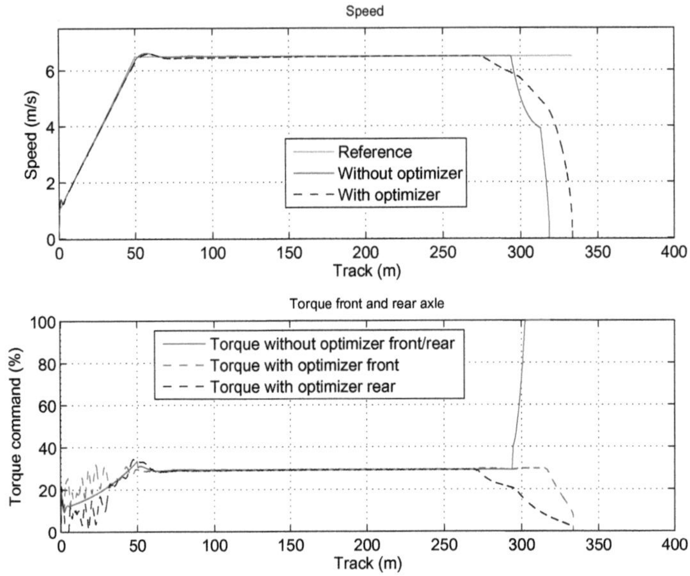

Figure 8.9.: Speed and torque profiles during straight driving while considering the temperature and State of Charge optimization goals, MiniMax

limiting the total drive torque. But, the difficulties of the optimizer to address the quickly changing and high slips (longitudinal slips higher than 25%) during the strong acceleration at the start become obvious in the torque plot (bottom of Fig 8.10). For such scenarios, a low-level slip controller should be added. The temperature management of the optimizer will become more obvious in the following experiments where the vehicle behaves as anticipated by the optimizer. In parallel, the charge levels between the two batteries are balanced as long as the temperatures are uncritical (Fig. 8.10). As the temperatures approach 70 °C, the charge levels of the batteries can no longer be balanced. Only at the end of the ride (around 260 m), when the charge levels become "sufficiently" critical, the overall drive torque is slowly reduced to protect the components. Without the optimizer, the front motor is fully loaded after depleting the rear battery in an attempt by the virtual driver to maintain the reference speed. As a result, the motor overheats and would be damaged in a real vehicle.

Performance of the Optimizer on a Complex Test Track

To investigate the performance of the optimizer in a challenging scenario where multiple optimization goals have to be considered at the same time, simulated

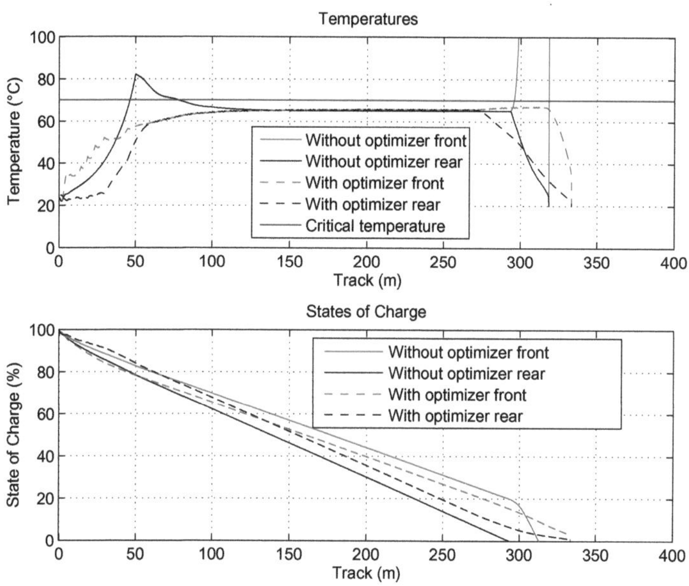

Figure 8.10.: Temperature levels of the drive motors and States of Charge of the batteries, MiniMax

test runs along the introduced complex track are analyzed. Figure 8.11 shows the tire wear and the temperatures recorded while driving with the optimized vehicle and a vehicle with an Ackermann steering and a constant torque distribution. As can be seen, the optimizer (MiniMax) balances the tire wear well and keeps the temperatures below the critical limits. Small wear fluctuations during driving are introduced due to the limited execution frequency of the optimizer and the simplifications in the prediction algorithms that, e.g., assume constant driver inputs.

Table 8.3 outlines several benchmarking factors calculated for the track[11]. Extending the already known influences, the average speed ($\bar{v}$), the average absolute lateral displacement of the vehicle ($|\bar{d}|$) from the reference trajectory and the average temperature of the drive motors and the power electronics is given ($\bar{T}$). Based on Tab. 8.3 and Fig. 8.11, the important influences on the optimizer are discussed in more detail for the example scenario.

The total wear generated by all versions of the optimization algorithm remains basically identical to the one generated by the Ackermann steering vehicle. Slight reductions result mainly from the more conservative accelerations of the optimized system to limit temperatures and the resulting slightly lower speed at critical points Total wear

[11]To reduce the impact of the initial vehicle launch, the values in the table were derived after the vehicle went twice around the track.

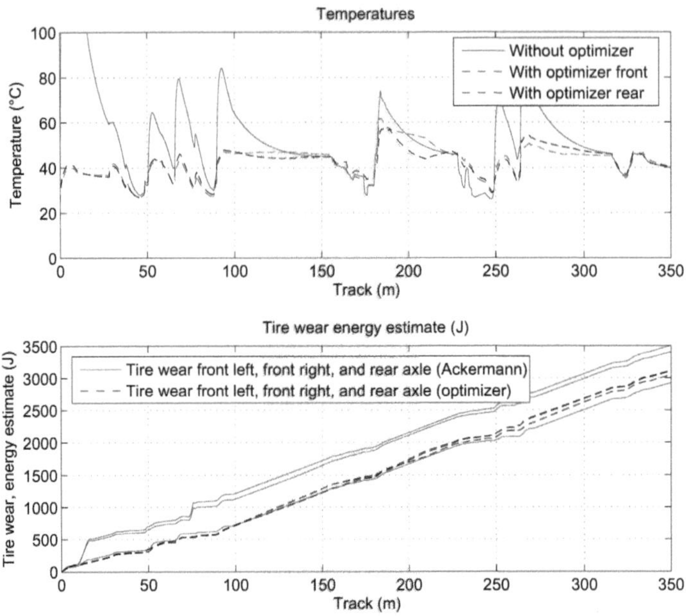

Figure 8.11.: Benchmark of the optimizer along a complex track

around the track. The evolutionary algorithm generates a total wear between 96% and 105% depending on the selection strategy from the pareto frontier. The values given in Tab. 8.3 result from a selection strategy that focuses on balancing of tire wear and obeying the torque reference. The most significant contribution to the reduction of tire wear in the optimized mode stems from the first left turn at the beginning of the track (at about 15m). Because of the huge longitudinal slips due to hard acceleration of the virtual driver with the unoptimized system at the beginning of the track, the side slip values are high at the first turn of the track and thus wear at the front tires increases significantly[12](Fig. 8.11). During the remaining drive, such influences diminish, and the average speed is hardly effected. It has to be pointed out that for some track geometries, the optimizer can also generate slightly increased wear. This especially occurs during dynamic driving with quickly changing steering angles or torques. Then, the execution intervals and prediction times by the optimizer are too long. This effect became obvious to some extent at the beginning of the straight segment test drive in the torque graphs (Fig. 8.10, bottom). Detailed evaluations on this matter will require a real human driver inside the vehicle to generate realistic inputs. If needed, a dynamic

[12]Longitudinal wear almost equal for both cases; consider also footnote 9 at the page 131.

Table 8.3.: Ackermann steering and equal torque distribution vs. optimized driving with MAX

Configuration	$\sum_{\forall i} w(i)$	$\bar{\sigma}$	$\bar{\sigma}_{\text{front}}$	$\hat{W}$	$\bar{v}$	$\lvert\bar{d}\rvert$	$\bar{T}$
Ackermann	100%	100%	100%	100%	100%	100%	100%
Opt. MiniMax	98%	7%	16%	91%	100%	59%	87%
Opt. GoalProgr	98%	5%	20%	91%	100%	58%	87%
Opt. evolutionary	103%	3%	15%	95%	100%	75%	98%
Ack (+10%)	103%	143%	496%	110%	100%	100%	100%
Opt. MiniMax (+10%)	101%	11%	52%	94%	100%	56%	88%
Ackermann (offset)	101%	124%	328%	104%	100%	100%	100%
Opt. MiniMax (offset)	100%	10%	15%	93%	100%	66%	89%

$\sum_{\forall i} w(i)$: total wear, $\bar{\sigma}$: standard deviation of wear for all tires, $\bar{\sigma}_{\text{front}}$: standard deviation of wear for the front tires, $\hat{W}$: maximal wear of a component, $\bar{v}$: average vehicle speed, $|\bar{d}|$: average absolute lateral displacement $\bar{T}$: average temperature of the drive motors.

adaptation of the prediction interval might be a possible solution, or the optimizer is turned off during such maneuvers. The same effect mentioned for the total wear holds true for the "improvement" in absolute lateral displacement from the reference line when driving the optimized vehicle. Thus, if temperatures are not considered, the lateral displacement generated by the optimized systems ranges from 93% to 95%. For all drives, the driver inputs remain almost unchanged for the unoptimized and the optimized system. Driver inputs

The considerable reduction in mean standard deviation of tire wear demonstrates the advantages of the online adaptation to changing conditions performed by the optimizer. The same observation has already been made during the real test drives with MAX, but in simulation the reductions are even stronger. Both the longer track in simulation providing the optimizer with more time to balance wear and the less noisy signals, which support clearer decision making, play an important role. When comparing the different optimization approaches for the simulated test run, the evolutionary algorithm provides the best performance. The tire wear is almost identical, and the remaining deviations stem most likely from the limited execution frequency and inaccuracies of the according predictions. Nevertheless, the design of the selection function to choose solutions from the pareto frontier is challenging. The current selection function sums up the fitness of a solution with regard to each goal and additionally weights the goals targeting at the "worst case" components. The weighting is similar to the dynamic weighting approach introduced for the evaluation of the optimization goals during gradient based operation. When the performance of the Goal Programming approach is analyzed, especially the dis- Standard deviation of wear crepancy between the performance in simulation and during real test runs becomes obvious. This observation strengthens the hypothesis that the Goal Programming approach suffers more from noisy data than the MiniMax approach.

Peak wear and temperatures

Similar to the results on MAX, the peak wear is reduced by almost 10% primarily due to balancing. Without considering the temperatures and thus sharper accelerations and almost identical total wear for all configurations, the peak wear is still reduced by 6% to 8% by the optimizers. The temperatures are kept below the limits by all optimizers, and the average temperature is reduced. The temperature reduction by the evolutionary approach is smaller. This is caused by the approach to select solutions from the pareto frontier, which focuses on wear balancing. Temperatures are only considered if they are expected to exceed the maximal temperature, which deviates from the dynamic weighting approach introduced for the gradient based design. If temperatures up to the upper limit are fully acceptable, the solution of the evolutionary algorithm is "more optimal".

Failure scenarios

The results from the complex test track confirm the trend seen in the previous tests. Due to the execution timings and the good results in both the real tests and the simulation, the MiniMax approach is chosen for a brief further investigation to derive the capabilities of the optimizer to handle unequal wear of components.

Unequal wear

In a first scenario, the wear generated at the most loaded tire was scaled by 1.1, which continuously adds an extra 10% of wear at that tire. In a real vehicle such increased wear may, e.g., be caused by wrong tire pressure. As a result, $\hat{W}$ for the Ackermann configuration increases by 10% as expected. Accordingly, the standard deviations and the overall wear increase. The optimizer is able to adapt to the changed situation and limits wear increase of the worst loaded component while still keeping the standard deviations low and fulfilling all other constraints. Similarly, the optimizer handles a second scenario with a constant wear offset of approx. 1.5% of the overall wear generated at all tires being added to the most loaded tire.

Summary for complex track

In summary, the optimizer operates nicely on the test track. In real scenarios, the short term effects would probably not have been as significant, but the long term advantage over an extended period of time can be assumed to pay off. The optimizer especially plays off advantages when dynamic situations have to be considered that require adaptations due to accelerations, resulting weight shift, or latent failures. In all configurations, the optimizer obeys the set constraints. The limitations due to the maximal execution frequency of the optimizer and the simplifications made for prediction of the temperatures and the wear are acceptable for normal driving and may even become less important if tasks with lower time-criticality are addressed.

For the evolutionary optimizer, the selection of appropriate solutions from the pareto frontier turns out to be the biggest challenge to generate good optimization results. Whereas, the multi-start cycles introduced for the gradient based system manage to handle the problem of local minima. With this set-up, the gradient based system performs almost as well as the evolutionary algorithm.

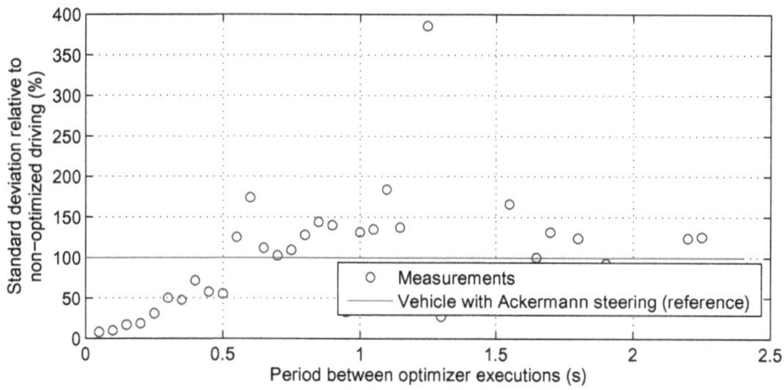

Figure 8.12.: Standard deviation of the tire wear depending on the execution frequency of the optimizer (SQP) relative to non-optimized operation

Execution frequency

The architecture introduced in Sec. 8.1.2 improves independence from the execution frequency of the optimizer. Still, the time criticality of the demonstration application – especially the optimization of tire wear – demands a certain minimal execution frequency. Figure 8.12 illustrates the standard deviations of the tire wear values for the already known complex track driven with different execution frequencies of the optimization system. It can be seen that at high execution frequencies the quality of results almost linearly increases with reduced time periods between executions of the optimizer. At time intervals, shorter than 0.2s, the quality starts to saturate, which is reasonable due to the timing constraints by the application. Up to a cycle time of approx. 0.4s the optimizer features better performance than an unoptimized vehicle with Ackermann steering. At lower frequencies, the performance decreases and becomes strongly dependent on the driven trajectory. Still, the virtual driver could successfully complete the track for all given frequencies without significant changes in the steering or throttle control. This indicates that the introduced system architecture, which keeps the optimizer off the time-critical path, enables the application-driven configuration of the system timings and even allows to use off-the-shelf optimization algorithms with high and varying execution times as a part of the outlined safety-critical application.

8.1.6. Conclusion

This section proposed a concept for cross-domain optimization of wear and load in highly flexible experimental vehicles. To achieve this goal, the developed system relies on an architecture that allows to combine off-the-shell optimization algorithms and estimation formulas to generate an optimal strategy that exploits the capabilities of the available actuators. As shown in experiment, the main goal of the optimizer to reduce the failure rates of components is achieved by avoiding unequal wear levels and preventing an overloading of individual units. This can then contribute to reduce maintenance effort, increase reliability, and then possibly strengthen the driver's confidence in the vehicle.

Optimization Strategy

The Multi Object Optimization approach chosen for the demonstration application allows to easily add, exchange, or modify criteria without having to consider the complete remaining system. The decoupling of the optimization task from the main vehicle control allows to adjust the execution frequency of the optimizer and facilitates the failure handling. Additionally, the dynamic weighting approach enables smooth adaptation of the optimizer to the upcoming critical situations.

Limitations of the system

Apart from these advantages, important downsides and limitations of the system have to be pointed out. For the presented demonstration application, the required computational power by the optimizer is high when compared to other typical systems within an onboard network. Still, the application was deliberately chosen to impose high demands on the execution times and to demonstrate technical feasibility even with off-the-shelf components. In terms of the specific application, the importance of a sufficiently good online estimation of critical wear and load values has to be stressed. Although the approximations made by the optimizer for wear prediction seem viable for the research context, the application in real vehicles will most likely require to improve the estimation formulas. Also, with appropriate measurement equipment, it has to be determined, whether short term influences on tire wear need to be captured or can be neglected as done in this thesis. Learning from the experiences made in the project, the normalization of the optimization criteria is vital to ensure a deterministic weighting of the optimization goals. The unnormalized values can lead to bad results in particular if the gradient based solvers are used. In summary, the presented system cannot replace basic protection mechanisms for the individual components but extends the existing systems by focusing on the dependencies across the vehicle.

Possible further developments

Further investigation of the optimization system should be carried out with a full-scale vehicle as MOBILE on a real test track. Especially, the perception of the handling of the vehicle by the driver in the optimized mode can hardly be derived from simulation or measurements with the scaled vehicle. If the good results presented in this work can be confirmed, further optimization criteria could be integrated into the system to consider other power consumers and actuators in the vehicle. Also, the application could be migrated from the comparatively "short-term" optimization of wear towards long-term optimization over vehicle life time, which will most likely still provide huge benefits for the driver but at the same time reduce the computational effort significantly due to the lower execution frequencies. Information from other traffic participants or road side units via Car2X, as considered by Saust [2014], could furthermore improve the quality of the optimization results.

8.2. Towards a Self-Representation for Vehicles

In Sec. 7.1 and Sec. 8.1, decision making systems have been introduced as a part of the strategic and tactical measures, which enable the proposed architectural approach for MOBILE. These systems derive decisions based on mathematical models and expert knowledge. This section starts from the question what input information is needed for decision making systems in general. As it turns out, the knowledge of the deciding system – the vehicle – about itself is a key factor for the proper decision making. For partially or fully automated[13] vehicles, Maurer [2000] derives that only if the vehicle considers its own state appropriately, it can derive the right decisions for safe driving that suite the current overall (traffic) situation. Thus, knowledge about the self is a vital prerequisite for automated or autonomous driving and focuses on the "inner" of the vehicle as opposed to other aspects mainly related to the environment. The challenge to represent this knowledge increases for highly flexible vehicles as MOBILE due to the higher number of degrees of freedom compared to conventional vehicles. Still, most research projects do not explicitly cover the knowledge about the self but only include certain aspects implicitly as hard coded and static decision thresholds or directives into the decision making algorithm. This makes it hard or impossible to adapt the system to the online changing conditions, e.g., a degraded operation, and to systematically consider the capabilities of the vehicle as a basis for its decisions. To address this challenge, this section introduces a more general and dedicated approach to self-representation for vehicles. The resulting system pools the knowledge of the vehicle about itself, adapts it to changes, performs targeted abstraction, and provides the knowledge to the driver and other top-level decision making systems. To design the self-representation system, the appropriate terminology is derived, which then inspires the definition of the system architecture.

8.2.1. Terminology

When defining ways to describe the knowledge of a system about itself, it seems reasonable to look at humans, who are usually well aware of their self and influence their behavior according to their abilities and skills. Consequently, the introduced terminology mostly stems from human sciences. For humans, the outlined knowledge about the self is referred to as "self-concept" (in German: "Selbstkonzept", compare, e.g., Boeck et al. [2009] and Rosenberg [1985]). Boeck et al. [2009] define the self-concept as follows[14]: *the self concept contains stable assumptions on the self and its abilities and skills* [Boeck et al., 2009, p. 916][15]. A similar definition can be found from Gerrig and Zimbardo [2002]. Gerrig and Zimbardo [2002] refer to

Self-concept

[13]Gasser et al. [2012] provide a detailed discussion of different degrees of automation in vehicles including example systems.

[14]Based on other literature, this definition is a severe simplification excluding social and other psychological aspects but will mostly suffice for application in the technical domain.

[15]Translation by the author.

the person's assumptions and knowledge about his/herself as the "person's mental model" [Gerrig and Zimbardo, 2002, p. G-12], which transfers well to the models of relevant components used in the technical domain to describe a system (compare also "Selbstmodell" (German, translation by the author: "model of the self") by [Metzinger, 1999, p. 158]). A more detailed assessment of the self-concept reveals that not only different topics, such as the skills or abilities, have to be considered, but also timely aspects are of relevance. Lipka and Brinthaupt [1992] summarize the work of several other researchers and subdivide the self-concept in a "baseline" and a "barometric" self-concept according to Rosenberg [1985]. The two categories refer to the mostly constant and the situation depended aspects of the self concept. For both categories, not only the mental development and the changes in the environment have to be considered, but also the person's physical changes have to be taken into account [Metzinger, 1999, p. 153]. Analogously, such timely aspects will have to be considered for technical systems.

Self-esteem

If a person evaluates his/her self-concept – basically he or she is thinking about him- or herself – the self-esteem (in German: "Selbstwertgefühl" compare [Daig, 2006, p. 34]) results [Mruk, 2006, p. 18, p. 206; Lipka and Brinthaupt, 1992, p. 66]. Consequently, the self-esteem and the self-concept are closely linked and influence each other. Some researchers, such as Rosenberg [1985], even regard the self-esteem as a facet of the self-concept. Negative "evaluation results" may cause the person to decide to build up new skills in order to improve his/her self-concept and therefore (hopefully) also self-esteem [Lipka and Brinthaupt, 1992, p. 75]. Still, it must not be forgotten, that the self-esteem does not only depend on the self-concept, but it is significantly influenced by external aspects, such as the social relations to others [Mruk, 2006; Gerrig and Zimbardo, 2002]. In an approach to scientifically capture a person's self-esteem, Rosenberg [1965] introduced the Rosenberg Self-Esteem Scale (RSES), which is a 10 item questionnaire provided to the evaluated person. This scale is now widely accepted and used also across different nations in different languages [Schmitt and Allik, 2005]. Other similar approaches are summarized by Lipka and Brinthaupt [1992]. To some extent, the self-esteem will also have to be covered in the technical system implicitly or explicitly, because this is what actually influences the behavior[16] of the system. Obviously, the intention will be to perform the evaluation of the self-concept as objective as possible in order to derive reliable and reproducible results.

It shall be pointed out that for the technical system this thesis focuses on abilities, skills, and a reduced set of technically motivated properties as part of the self-concept and the self-esteem. Other factors, such as the emotions and feelings that are included for humans, are not transferred to the technical domain. Additionally, the self-concept as referenced in psychology is always personally biased and does not necessarily reflect the real abilities, skills, or attributes[17]. For the technical

[16]Behavior refers to "the action, reaction, or functioning of a system, under normal or specified circumstances" [Collins, 2013].

[17]An attribute represents a "property, quality, or feature belonging to or representative of a person or thing" [Collins, 2013].

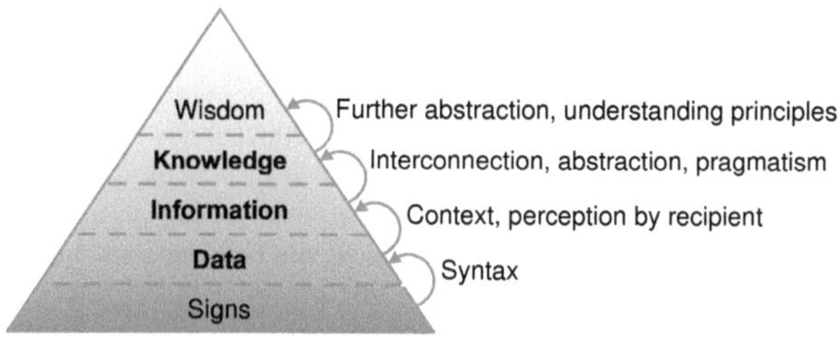

Figure 8.13.: From signs to wisdom according to [Voß and Gutenschwager, 2001, p. 13] and [Bodendorf, 2006, p. 1]

system, strong personal bias, introduced, e.g., by the developer, should be avoided, although at some point it might be desirable that the vehicle has its own bias driven by its own skill and ability set.

Self-representation

The expression self-representation (in German: "Selbstrepräsentation", Metzinger [1999]) is frequently used synonymously to self-concept [Ruel, 1987]. Despite this partially synonymous usage, Ruel [1987] argues that *the self-representation rather represents the combination of the self-concept and the self-esteem*. If the self-esteem again is assumed to be part of the self-concept as done by Rosenberg [1985], self-concept and self-representation turn out to be synonymous again. For the approach presented in this section, the two aspects self-concept and self-esteem have to be covered, which goes along well with both wording approaches. For clarity, this section follows the definition of Ruel [1987] for the self-representation and refers to the self-concept and the self-esteem if individual aspects are referenced.

Further definitions

Accepting the above definitions, the key aspects "knowledge", "ability" and "skill", which are referred to in the definitions, shall be examined. To start with, it is clarified what knowledge refers to. This is of special interest, because the vehicle has to be provided with the knowledge about its abilities and skills, which again has to be stored in some form. As the resulting technical system has to connect humans rather soft perception of things with the technical facts, definitions referenced here are mostly taken from the field of industrial management. This field of research, amongst others, focuses on the knowledge management in companies and provides suitable definitions that connect the perception of a system by humans with the computational data processing. In doing so, the relation between data, information, and knowledge is of special interest. Figure 8.13 summarizes these relations, which are detailed in the following.

Data

Starting from the bottom, *data is perceived as isolated facts that are per se not related to any use case, purpose, or meaning* [Hoffmann, 2010; Zack, 1999 according to Gerstlauer, 2004, p. 9]. Data is represented by a set of signs that are connected by syntactic rules [Bodendorf, 2006, p. 1], e.g., the signs '1', '2', and 'V' can be connected to '12V'. This step from signs to data is accepted by almost all related research projects that were examined.

Information

Definition of terms becomes more challenging when the step towards information is taken. As indicated by the gray shaded pyramid, the transitions between the different stages are fluent and there is no general agreement on a hard separation between data and information (see, e.g., Spitta [2007] for a critical discussion of different aspects). Still, commonly agreed is that *information – other than data – includes the current situation and the receiver's personal background for interpreting data.* Therefore, data becomes meaningful to the receiver and thus information [Davis and Olson, 1985, p. 200; Hoffmann, 2010, p. 5; Willke, 2004, p. 31]. As a result, data is merely used as a mean for "communication, interpretation or processing" [ISO/IEC 2382-1, 1993, p. 6]. For example, the data element '12V' only makes sense to the receiver if he or she can associate 'V' with voltage, and relate the value to the current situation and the examined system[18]. As pointed out by Floridi [2005] this receiver-orientation of information comes along with a challenge: depending on the receiver, data can remain data or transition into varying information. A person with no technical background in general or missing background on the analyzed system might not associate anything with '12V' and thus data won't become the intended information. Still, non-primary information, such as the information "missing documentation", can be contained in such data [Floridi, 2005]. For this work, only the primary information is considered. In general, this discussion of receiver-orientation of information strengthens the idea of a skill and ability based interface to the user as outlined by Maurer [2000] and followed in this section, because typically every user of a vehicle will be aware of basic driving tasks and thus be able to interpret related indications appropriately.

Knowledge

Towards knowledge, the clarity of nomenclature continues to blur. Thus, this paragraph provides only a brief assessment of a minimal core of commonalities in the definitions. In general, *knowledge is perceived as a more condensed abstracted and interconnected view on information with respect to a certain problem* [Bodendorf, 2006; Voß and Gutenschwager, 2001]. Therefore, knowledge is mainly attributed as "pragmatic". ISO/IEC 2382-1 considers a knowledge base as "a database that contains inference rules and information about human experience and expertise in a domain" [ISO/IEC 2382-1, 1993, p. 22], which also stresses the strongly interconnected character of knowledge. Voß and Gutenschwager [2001] add "wisdom" as an additional layer on top of knowledge, which then focuses even more on the understanding of principles underlying an observed process or state and thus features a further condensed character. Wisdom is not considered for the technical system, because it is hard to distinguish from knowledge and not relevant for the application. Summarizing the gathered definitions from data to knowledge, two general facts are important for the discussions in this section: from data to knowledge, (a) a targeted "pragmatic" process of abstraction is needed that includes increasingly more background information, and (b) towards knowledge the number of interconnections between the individual data sets increases strongly.

[18] The difference between data and information can also be found in the definitions of ISO/IEC 2382-1. The "interpretation of data" leads from data to information [ISO/IEC 2382-1, 1993, p. 25].

Abilities and skills

Concluding the stepwise definition of important terms, abilities and skills are defined. Incorrectly, ability and skill are frequently not distinguished, such as in common parlance or even in more general lexica. The definitions of skill and ability are again inspired by human sciences, more precisely by the fields of sports science and psychology. In these fields, a clear distinction between abilities and skills is a matter of ongoing research. Bergholz [2003] provides a detailed discussion of international work defining abilities and skills. In particular, the difficulty to derive a common definition of ability becomes obvious. Carroll [1993] puts it in a nutshell when he states: "Although the term ability is in common usage both in everyday talk and in scientific discussions among psychologists, educators and other specialists, its precise definition is seldom explicated or even considered. [...] Oddly enough, dictionaries are of little help in developing an exact, analyzed meaning of the term" [Carroll, 1993, p. 3]. *A skill, however, is commonly accepted as being directly linked to an observable (part of an) action.* For example, "running" is a base skill, which can directly be observed by others if the person runs [Bös, 2003, p. 2]. Starting from there, various strategies are followed to define abilities. This section only presents a reduced set of selected commonalities. Almost all *researchers perceive abilities as the basis for being able to learn and use a skill.* Carroll [1993] compares a person's level of ability with the "maximal performance" or the chance of a person to perform well. This perception is adopted for use in this section. Thus, a person or system not having a sufficient level of ability for a certain task will never be able to build up the skill to perform the task – unless he/she/it manages to build up new abilities, which is possible to some extent, e.g., by training or by additional equipment. The other way round, a person or system having the ability to perform a given task does not necessarily also have the skill. For example, having legs does not necessarily give a system or person the skill to walk. The quality of an action carried out by a person or system depends on the one side on the level of ability (e.g., how strong are your legs) as a prerequisite, and on the other side on the level of skill (e.g., have you been trained as a professional runner) earned by experience and practicing. If different types of ability are assessed in more detail, one can distinguish the motor abilities, such as power, endurance, speed, motility, or coordination [Bös, 2003, p. 2], and cognitive abilities that are related to cognitive tasks which especially focus on "processing of mental information" [Carroll, 1993, p. 10]. A simple example task that relies mainly on cognitive abilities would be to memorize and repeat a series of numbers [Carroll, 1993, p. 10]. Several research projects distinguish these two basic types of ability and investigate their mutual dependencies to, e.g., improve the education of children [Jansen et al., 2010; Voelcker-Rehage, 2005]. More details on the definition of skill and ability, which exceed the condensed approach in this thesis, can be found in the already referenced work by [Bergholz, 2003, pp. 74] but also in [Bös, 2003, p. 2]. Analyses with a focus on the acquisition of skills by learning and based on the individuals abilities are presented by [Eberspächer, 1987, pp. 178, pp. 474] and [Schnabel et al., 2011, pp. 136], which lead further into the approaches followed in sports science.

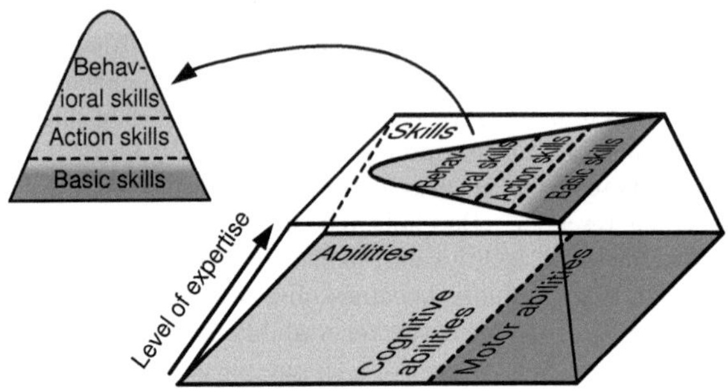

Figure 8.14.: Relating the skill hierarchy and abilities

Transfer to a vehicle

Transferred to a vehicle, skills are the observable actions carried out by the vehicle and thus are vital for the decision making system. For the vehicle, skills will mostly depend on the algorithms implemented on the vehicle to control available actuators or evaluate sensors.

Hierarchical abstraction

For abstraction in the technical system, this thesis organizes the skills of the vehicle hierarchically. In sports science, researchers refer to more complex actions that include several basic skills, such as dribbling in basket ball, as complex motor performance skills (translated from German "komplexe sportmotorische Fertigkeiten" [Bös, 2003, p. 186]) and thus build up a "skill hierarchy" starting from the most basic skills as running. Deviating from this nomenclature, because the wording only supports two hierarchical levels, this work adopts the nomenclature introduced by Siedersberger [2003] that was developed for a hierarchy of abilities associated to a highly automated vehicle. Siedersberger [2003] distinguishes three levels of abilities: a skill, an action, and a behavioral level. Coinciding with the already introduced definitions, elements at higher layers consist of multiple elements from lower layers, and the degree of abstraction increases towards the highest layer. As this work focuses on representation of observable actions carried out by the vehicle, the nomenclature by Siedersberger [2003] is adapted from abilities to skills. Abilities, as the enabling elements for skills, will be implicitly modeled to some extent in the system by the performance limits of the vehicle. The following skill hierarchy results: basic skills, action skills, and behavioral skills. Figure 8.14 summarizes this wording convention in the shown pyramid. As indicated by the coloring, basic skills are significantly influenced by motor abilities. Still, even the simplest task requires cognitive parts. Starting with action skills, the appropriate combination of basic skills becomes important, which leads to an increasing effect of the cognitive abilities on the resulting final and observable performance. The figure furthermore references the current "level of expertise". The higher the level of expertise for a

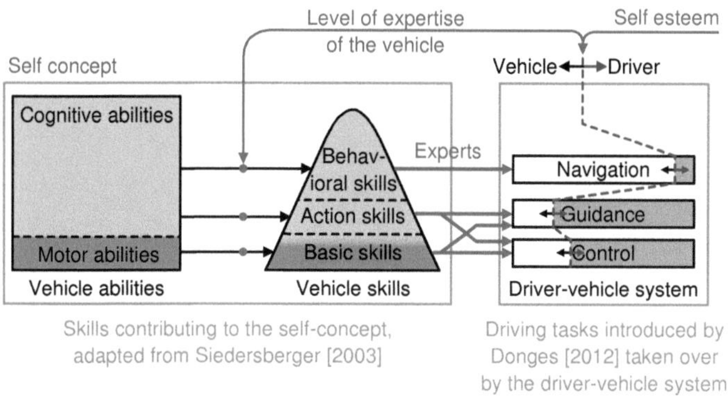

Figure 8.15.: Relating the self-concept to the three layered model of Donges [2012]

given skill is, the more of the potential provided by the abilities of the system can be transferred into skills and exploited for vehicle operation. The nomenclature again leans on the definition by Siedersberger [2003], who associates an "expert" to each ability for execution of the related action. The larger "area" of the ability section compared to the skill pyramid in Fig. 8.14 symbolizes that usually it will not be possible to fully exploit all available abilities by skills. For the abilities, no hierarchy is introduced in this section, but could be if required by an application. The given separation between ability and skill is strongly reflected in flexible vehicles as MOBILEbut also in automated vehicles as will be introduced.

Relating results

Concluding the introduction of the skill hierarchy proposed in this work and the related terminology, this section puts the results into relation to existing research by Donges [2012] and Rasmussen [1983]. Donges [2012] presents a three layered model of the driving task. Due to the task/result oriented design approach[19], the model contains the three layers navigation, guidance, and control that are derived from the actions that have to be performed by the driver of a vehicle. Figure 8.15 relates this model with the introduced structure and terminology for the self-representation. It can be seen that the driver or an automated control algorithm can utilize the available skills. For the driver, the current "skill state" is informative and can contribute to safe control of the vehicle, whereas a control algorithm can directly exploit the provided knowledge. In a partially automated system, the level of expertise in combination with the results of the permanent self evaluation of the vehicle (self-esteem) determine which tasks can be performed by the vehicle and which ones have to be carried out by the driver – relating to the "degree of automation"[20] introduced by Maurer [2000]. As both the structure introduced by Siedersberger [2003], which was taken as a main reference for this work, and the

[19]German: "aufgaben-/ergebnisorientiertes Modell"
[20]Translation from German: "Grad der Automatisierung"

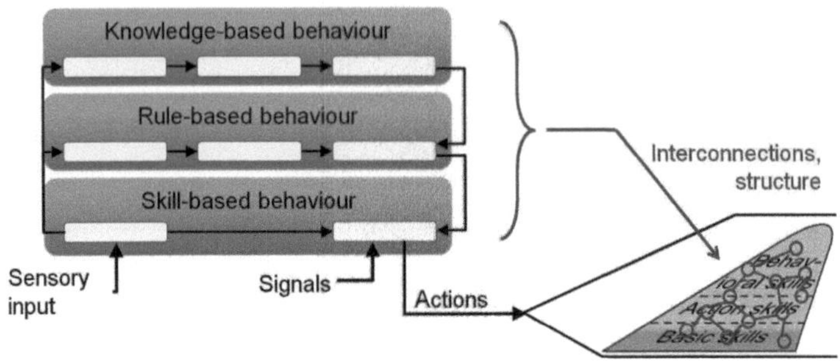

Figure 8.16.: Relating the skill hierarchy to the model of human behavior by Rasmussen [1983]

approach chosen by Donges [2012] follow a task oriented design, both structures combine well. The number and structure of the arrows in Fig. 8.15 between the hierarchy levels of the vehicle skills and the levels of control by Donges [2012] are a proposal and have to be evaluated for each individual case. By tendency, the corresponding hierarchical levels match each other as will become obvious in Sec. 8.2.4 when the demonstration application is discussed.

Rasmussen [1983] introduces a general model to describe the human approach to "act"[21] when confronted with a task. Thus, he is following a fundamentally different approach than Donges [2012], and a model results that is less domain specific and applicable to multiple different fields involving human operators. Rasmussen [1983] defines from bottom to top-level "skill-based behavior", "rule-based behavior", and "knowledge-based behavior" for human actors. The skill based layer contains elementary "sensory-motor" [Rasmussen, 1983, p. 2] actions that are then combined into more complex actions according to the rules stored at the second layer for rule-based behavior. At the top level, knowledge based behavior is activated if the human operator is confronted with a so far unknown situation for which no stored rules exist. If the decision making system in the vehicle is structured according to the approach by Rasmussen, the skills in the skill hierarchy proposed in this section are the observable outputs after a given sequence of basic skills has been executed (Fig. 8.16). The complex skill sequences result from a rule stored at the second layer of Rasmussen [1983] and accordingly included basic skills. Thus, the skill hierarchy captures the results of the decision making process given by Rasmussen [1983]. If the performance of each relevant skill combination is stored in the self-concept or can be predicted, a decision making system structured according to Rasmussen's model can take this as a basis for the selection of the best suited rules. In summary, both models by Rasmussen [1983] and Donges [2012] with their differing foci can be used to structure a decision making system or analyze the human behavior [Damböck et al., 2012]. The presented self-concept can be combined with both

[21] German: "vorgehensorientiertes Modell"

approaches, and the nomenclature remains valid.

Summary of Contributions

So far, a structure for a novel self-representation system for technical applications derived from a survey and the targeted abstraction of terminology has been introduced and related to existing works in similar fields. The system is mainly based on skills and abilities, which then form the input to self-evaluation and decision making systems. Starting from these fundamental considerations, the following section will detail a hierarchically abstracted self-concept for vehicles. This self-concept is continuously updated and adapts to the current situation to take into account timely aspects. Thus, the system supports reconfiguration and adaptation. In MOBILE, the knowledge derived from the self-concept is used for the driver information, preventive safety measures, and to trigger upgrade scenarios. Further usage for fully or partially automated vehicles is referenced in the outlook.

8.2.2. Self-Concept and Self-Esteem for Vehicles

To start the development of a technical system forming the self-concept, an overview of similar research approaches that have already been followed for flexible or automated vehicles is given. This section focuses on research projects that explicitly integrate some dedicated (software) system to cover the abilities, skills, and properties of the vehicle. Other planning and decision making systems that include this knowledge implicitly, which is in any case mandatory [Siedersberger, 2003, p. 74], are neglected.

Universität der Bundeswehr München

The work of the research group of Dickmanns at the Universität der Bundeswehr München forms the basis for the investigations in the MOBILE project. The contributions of Maurer [2000] describing the overall concept for setting up a flexibly automated vehicle with machine perception, and the works of Pellkofer [2003] and Siedersberger [2003] are considered, who investigate the ability-based decision making and vehicle control as a part of the framework introduced by Maurer [2000]. Within the research project, the knowledge of the vehicle about itself is referred to as "self-representation" (translated from German "Selbstrepräsentation" by the author). This self-representation includes, amongst others, abilities, the shape of the vehicle, its dynamics, the degree of automation, and, e.g., also the current optimization criteria that are applied to execute a maneuver [Maurer, 2000, p. 59]. Thus, the chosen approach to self-representation correlates well with the introduced definitions in this section. As a core part of the self-representation, a hierarchical net of abilities is introduced that covers all relevant abilities needed for automated driving within a defined set of domains[22]. For the abilities, the already introduced three hierarchical layers were defined: "skill level", "action level", and "behavior level"[23] [Siedersberger, 2003, p. 75]. The ability net reflects the static dependen-

[22]According to Maurer [2000], a domain defines the general features of the environment of the vehicle including the relevant rules for this environment. For example, a highway is a highly regulated domain [Maurer, 2000, p. 4].

[23]Translated from German: "Fertigkeiten", "Handlungsfähigkeiten", and "Verhaltensfähigkeiten"

cies among the abilities and the dynamic availability of the abilities. The latter aspect addresses any subsystem failures or the quality of the currently available knowledge [Siedersberger, 2003, p. 48]. Within the network, the abilities are again categorized in abilities that relate to the perception system, the lateral, or the longitudinal dynamics. This separation is introduced at the lowest hierarchical level of abilities, because the individual abilities can clearly be associated to a certain set of actuators or sensors that belong to each of the categories. Towards the higher hierarchical layers, the separation vanishes. Each ability within the ability net is associated an expert that executes the action corresponding to the ability. This expert also monitors proper execution of the triggered actions and is able to predict the results of an action within, e.g., a certain time interval if requested. The prediction is needed by the decision making and planning modules of the automated system [Siedersberger, 2003, p. 90]. For the approach presented in this section, especially the hierarchical approach to organize abilities and the idea of experts responsible for executing the actions will be adopted. Still, a distinction between skills and abilities according to the definitions introduced in the previous section is missing and needs to be considered for the outlined flexible vehicles, as will become obvious later on.

Project "Stadtpilot"

The project "Stadtpilot" also conducted at the Institute of Control Engineering at the Technische Universität Braunschweig deals with the automated driving in Braunschweig city with the vehicle "Leonie". The project is referenced to outline the potential of a future merge between the parallelized projects Stadtpilot and MOBILE. Currently, the decision making and control system of Leonie includes knowledge about the ego vehicle. Nevertheless, the current "knowledge" of the vehicle about itself is strongly safety driven. As outlined by Reschka et al. [2012], the vehicle evaluates several performance criteria covering measurements from the odometry or the perception system. The measurements are combined by using heuristics. The resulting performance criteria can trigger functional degradation in the vehicle or even emergency maneuvers if the system partially fails. Implicitly, these criteria can be perceived as a quantification of certain skill levels, which are then considered for decision making. Still, the system does not yet represent a self-concept as introduced. Mainly due to the primarily safety driven background for low level control, the introduced measures are closer related to the low level action derivation system PFDH introduced in Sec. 7.1. Nevertheless, the work of Reschka et al. [2012] points out the need for detailed knowledge about the ego vehicle.

Project SPARC

Holzmann [2008] introduces the approach followed in the SPARC project to switch between a virtual and a real driver. In the project, the skills of the vehicle are not explicitly represented, but confidence values into the sensors and the agents executing certain actions are calculated similar to the monitored values in the project Stadtpilot. The confidence values can again be interpreted as measures for some kind of skill level for the related tasks. In parallel, the approach by Holzmann [2008] estimates the performance of the real driver online. The resulting performance levels of the virtual driver (the vehicle) and the real driver are compared and a Fuzzy Logic based system determines the priority of each driver. ("A confident driver can override the assistant system and vice versa" [Holzmann, 2008, p. 177]). The work of Holzmann [2008], though not representing skills or abilities explicitly, demonstrates the need for a detailed monitoring and evaluation of the skills of the vehicle especially in a partially automated vehicle if the driver and the technical system shall cooperate smoothly at varying degrees of automation.

Schneider [2010] investigates the modeling of driving situations for safety critical driver assistance systems. As the ego vehicle is a vital part of a driving situation, it needs to be included. Nevertheless, the work only regards static attributes and properties of the ego vehicle analogously to the strategy followed for other traffic participants. Thus, a part of the self-concept is considered, but the actions that can be performed by the vehicle or a degradation concept are not explicitly included although highly relevant for the decision making. In particular, in safety critical situations, a detailed understanding of the skills and performance level of the own vehicle is vital. BMW-Group

Reichard [2004] proposes a generic architecture suitable to add a system health representation to automated systems. His key motivation is that in "fully autonomous systems, a human operator may not be able to intervene or rescue the system" [Reichard, 2004, p. 2], and thus the system needs to monitor itself and recover if required. As core tasks, Reichard [2004] tries to derive the expected demands on critical components and subsystems during maneuvers and compares these requirements with the current subsystem or component health state. To address these tasks, a system monitor is added, which gathers data from decentralized components. By pragmatic relation of these pieces of information, knowledge about the health state of the system is generated, which is then referred to as self-awareness[24]. The health monitoring is not associated to a functionality or the abilities of the technical system but to individual hardware units following a strongly hardware centric approach. Still, the effects of detected faults are known by the central decision making system, and thus skills and abilities are implicitly represented there. A core part of the work is the decentralized acquisition of data with little overhead on the communication systems. Reichard [2004] recommends to share only preprocessed data that already features a high level of abstraction and thus consumes little bandwidth on communication systems. The so far outlined systems of other research groups feature a mostly centralized hardware architecture on one or a small number of computers that are connected with high-bandwidth bus systems. Thus, these projects did not have to face the challenges accompanying a distributed system architecture with computationally low power devices and a low-bandwidth data bus system as a part of the signal processing path. Applied Research Laboratory

[24]This does not comply with the definitions deduced in the MOBILE project. This work would refer to the mentioned aspects as a part of the self-concept. "Self-awareness" (In German "Selbstbewusstsein" [Kemmerling, 2000, p. 21]) is defined as "cognizance of the autobiographical character of personally experienced events" [Gerrig and Zimbardo, 2002, p. G-12]. For a more detailed and controversial discussion of self-awareness and underlying concepts refer to Metzinger [1999].

Autonomous robots

In the field of robotics, research on autonomously acting systems has long been ongoing. Long et al. [2007] provide an overview of architectural concepts including implementation details of several systems. Surprisingly, hardly any system takes aspects of an explicit self-concept into account. Knowledge about the ego robot is implicitly integrated into the system. Still, work in the field of learning robots highlights the need to distinguish between ability and skill. For example, Kurz [1994, 1995] introduces a robot that learns new skills in the field and is afterwards able to apply these skills in new scenarios. Thus, the robot was able to fulfill the associated tasks right from the start but was missing the skill to actually execute the required actions. The acquisition of new skills requires a way to structure and store skills. Frequently, an unstructured representation is relied on, e.g., with Neural Networks trained based on measurements. This way to represent skills is suitable for an autonomous robot in a highly unstructured environment with little regulations on its behavior. For vehicles in real traffic, this representation of knowledge tends to be too hard to evaluate for human experts when performing plausibility checks or adapting the system to a fixed set of rules, such as the public traffic regulations. Schröder [2009b] provides an evaluation of the machine learning approaches relative to classical decision making systems based on the vehicles participating in the DARPA Urban Challenge and confirms the limitations for the use in public traffic [Schröder, 2009b, p. 24].

Motor and perceptual schemas

Arkin [1989] introduced motor schemas to control an autonomous robot and the AuRA[25] architecture based on these schemas. The approach was inspired by neuro-scientific and psychological research and supplements each motor schema, such as "avoid obstacle" or "move ahead", by a perceptual schema that provides the necessary perception of the environment to set the parameters of the motor schema. Finally, the robot is controlled by multiple instantiations of the schemas that are overlaid. Therefore, a potential-field-like approach is introduced. Thus, the approach differs strongly from the approach by the research group of Dickmanns that clearly selects individual skills for given tasks. The schemas introduced by Arkin [1989] can be seen as the skills of the robot [Arkin, 1992, p. 118]. Still, the approach does not fully cover the knowledge of the robot about itself. In particular, the effects of over-actuation on the decision making of the robot have so far not been considered. Also, defects or degradation of the robot are not taken into account, although changing environmental conditions with negative effects on the robots handling are considered [Arkin and Balch, 1997]. The latter influences also have to be considered for vehicles, e.g., when different surface or weather conditions impact handling.

Cognitive automobiles

Schröder [2009b] introduces a decision making and trajectory generation system for "cognitive automobiles". The implemented three-layered behavior-based decision making system widely resembles the approach introduced by Pellkofer [2003] and Siedersberger [2003]: a system monitor performs plausibility checks and provides estimates of the expected quality of execution of an action. This task is asso-

[25]Autonomous Robot Architecture

ciated to the experts in the original structure by Pellkofer [2003] and Siedersberger [2003]. The successful re-use of the basic approach indicates good applicability for vehicles. Furthermore, Schröder [2009b] provides an overview of further research projects on decision making. Amongst others, the reactive architecture of Laugier is referenced [Laugier et al., 1998; Laugier and Fraichard, 2001]. This approach applies skills similar to the motor and perception schemas by Arkin [1989] to control an autonomous vehicle, but does not include an explicit definition or representation of abilities or skills.

Summary

In summary, few research groups are working in the field of self-concepts for flexible and automated vehicles. A possible reason for the lack of research may be that functional issues especially with regard to environmental perception in automated vehicles are still far from being solved, and thus the own vehicle is frequently assumed to work properly and its constant skills are implicitly represented. Still, Reichard [2004] points out that "autonomous system approaches have traditionally struggled with the representation of the external environment in which the system operates. An accurate representation of the internal state – including the health of critical subsystems – can be equally challenging" [Reichard, 2004, p. 1]. Another issue may be that highly flexible vehicles are not commonly available yet. Typically, vehicles with standard actuators for steering and drive with no couplings in the control of the actuators are used in the research projects for autonomous driving. Thus, the vehicle itself and its current set of skills are rarely considered. Still, even the performance of electric vehicles with standard actuator set-up can vary significantly depending on the charge status of the main battery or the temperatures of the drive motors. Flexible vehicles take another step ahead. As a result, the ego vehicle needs to be represented in some way, which triggered the development of the system introduced in this section. In doing so, the work of the research group of Dickmanns outlined by Maurer [2000], Siedersberger [2003], and Pellkofer [2003] provides the foundation. As will be shown, the system developed in the MOBILE project extends the existing structure in two main fields. First, the higher flexibility of the over-actuated vehicle including the influence of the electric drive train and the resulting effects on abilities and skills are taken into account, and second the interfacing to the human driver for maintenance and upgrade scenarios is considered. Therefore, an incentive system is added to support the evolution of the vehicle according to the needs of the driver. The core focus of the developed system is up to now rather driver information than automatic vehicle control.

8.2.3. The Approach to Self-Representation

The highly flexible vehicle MOBILE is both intended as a research tool and is itself a subject to research. Accordingly, the vehicle features multiple actuators providing capabilities that surpass the ones of series vehicles by far (Cha. 4). The vehicle is able to perform numerous maneuvers within the physical limits ranging from a typical front-wheel-steer driving to turning at a spot. The computational power available for control of these maneuvers is scalable due to an appropriate power supply and modularity in electronics. Currently, the ego-motion control of the vehicle is only limited by the available control and sensor data processing algorithms. Thus, the perceivable skill level of the vehicle deviates substantially from the ability-wise possible level. If additional sensors for environmental perception are added, which is done in ongoing research, the gap becomes even wider. As a result, an appropriate communication of the current skill set of the vehicle to the driver is vital, in order to ensure that the driver knows which skills he can rely on to what extent. Therefore, a pragmatic abstraction of the current system state starting from information about the vehicle and online data is performed. This knowledge about the (increasing) skill level not only informs the driver but also provides information on the current state of development to developers. Adding new functionality or improving existing modules increases the skill level of the vehicle and indicates progress to all participating researchers.

Requirements

To start the development of a system driven by the above needs, requirements are derived. Above all, the self-concept needs to describe the state of the ego vehicle. This knowledge shall be represented in a skill-based manner, which makes it more intuitive for the driver to use and to integrate into a decision making system for automated vehicle operation. Moreover, the provided knowledge shall contribute to functional safety and reliability, because critical driving situations can be avoided if an a-priori estimation of the planned maneuvers indicates that the currently available set of skills or the level of a certain skill do not suffice to perform the maneuvers safely. At the same time, the self-representation system should integrate into the EE system with little consumption of computational power and communication bandwidth. Taking furthermore the tooling character of the vehicle into account, a good usability not only for the driver but also for the developer implementing the self-concept has to be ensured. Thus, two important aspects have to be covered. (a) New extensions to the vehicle have to be easy to integrate into the self-concept by a modular software design, and (b) the interface to add knowledge to the self-concept, e.g., knowledge about the interdependencies or performance levels of components, has to be easy to use for the developer. These requirements for the self-concept directly relate to the top-level requirements for MOBILE as summarized in Table 8.4.

Solution strategy

To implement the skill based self-concept driven by the given requirements, a system comprised of two main modules is implemented: an information (data) base[26] and a system containing the knowledge[27] needed for the self-concept.

Information base

The information base gathers available measurements in the vehicle, associates the measurements to functional modules, and adds additional information to each measurement, such as the units, ranges, types, or descriptions. This additional information can be provided online during vehicle start-up or offline by a suitable configuration file → R1.S1. During operation, the information base does not require

[26]To clearly distinguish data and information, the following section will refer to the information (data) base as "information base", because it contains "meaningful data" for other technical systems or the experienced driver.

[27]In the following, this system will be referred to as knowledge base.

Table 8.4.: Requirements for the self-concept associated to the top-level requirements given in Tab. 4.1

R1	**Mechanical and electronic modularity**
R1.S1[a]	The self-concept shall facilitate easy exchange of hardware and software components and regard the over-actuation of the vehicle.
R1.S2	The complex hard- and software structure of the vehicle shall be encapsulated in a functional/skill-based representation.
R1.S3	The system shall support upgrades of the vehicle driven by the needs of the driver.
R2	**"Open-source" vehicle**
R2.S1	The system shall be easy to configure and provide a flexible interface for other applications in the vehicle.
R2.S2	Monitoring of the vehicle control system has to be supported.
R3	**Functional safety**
R3.S1	The system shall support the driver to avoid critical situations by providing information on the current and the expected system state.
R4	**Limited degree of hardware redundancies**
R4.S1	The system providing the self-concept shall integrate into the vehicle without generating communication overhead during operation. And, adding it to the vehicle must not violate any safety goals.

[a]R1.S1 stands for the first requirement on the self-concept derived from the top-level requirement R1.

the application modules to transmit any more data than needed for executing the application [→ R4.S1]. Based on the communication traffic, the stored data in the information base is updated in real-time. These pieces of information are then provided to other systems in the vehicle via an Ethernet-based query-response interface [→ R2.S1]. To configure the information base, the expert is supported by a graphical user interface [→ R2.S1].

Starting from the information provided by the information base, the knowledge base forms the self-concept based on the skills and properties of the vehicle [→ R1.S2]. Knowledge base In this process, the most important task is to transfer expert knowledge into the technical system. Frequently, Fuzzy Logic is used to perform such a task (see also Sec. 7.1.1), which is also the way followed in this work. The fuzzy system models the dependencies among different pieces of information to finally derive the skill hierarchy. Online, each skill is associated a performance level. These levels are

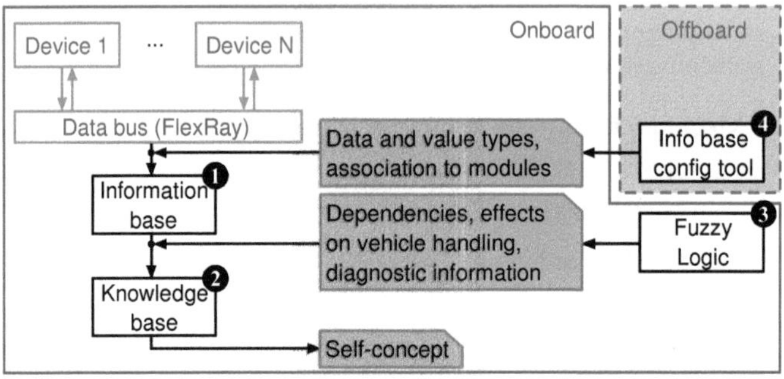

Figure 8.17.: Structure of the self-concept generation process

then taken as an input to determine whether a basic set of skills needed for driving is available. Every additional or improved skill exceeding the basic requirements causes the vehicle to be ranked better and vice versa. Additionally, diagnostic → R2.S2 and upgrade → R1.S3 information are added on each of the hierarchical skill layers. The driver can then access the current skill-set and -level and the diagnostic information. Additionally, he can access hints for possible upgrades of the vehicle to improve specific skills according to his needs. In case of severe system failures, a vehicle start can be prohibited or certain maneuvers are marked as unsafe → R3.S1.

Core contributions

Following this solution strategy, the system proposed in this section extends the developments in the related work by several key aspects. On the theoretical side, a clear definition of nomenclature and the resulting system structure adapted for the technical domain take research towards self-representation in vehicles a step further. Also, the evaluation of skills relative to a reference vehicle with basic handling characteristics comparable to the approach outlined for safety evaluation (Cha. 6 and Cha. 9) provides an understandable minimum set of skills and skill levels to the driver or a system for automatic control. The outputs of the diagnostic system, which are integrated into the self-concept, help to identify, handle, and solve occurring problems and are accompanied by a unique incentive system that indicates possible improvements of the vehicle in a way the driver desires or needs. For the developer, the modification of the self-concept is supported by a modular design and graphical tools to configure the system. The system itself is designed to operate with minimal consumption of communication bandwidth in the vehicle.

8.2.4. Implementation of the Self-Concept

Figure 8.17 outlines the most important components of the self-concept implemented for MOBILE. ❶ to ❹ in the figure indicate software modules. ❶ to ❸ are executed online, while ❹ refers to an offline configuration tool for setting up the

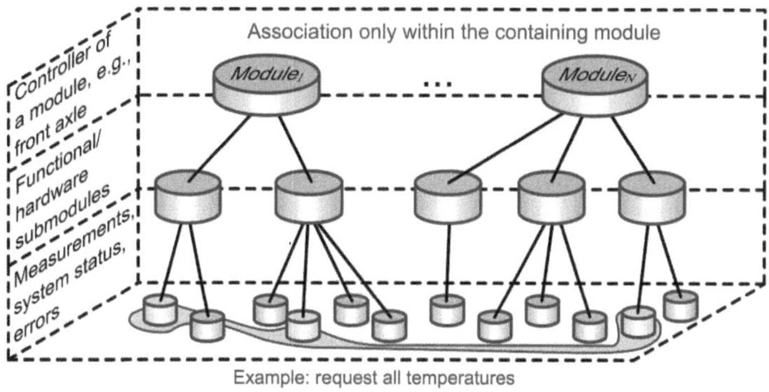

Figure 8.18.: Internal hierarchical structure of the information base

information base. The slanted rectangles indicate "work products" generated by a software module. The following section details the software modules starting with the information base.

Information base

The information base ❶ is supposed to gather the available data in the vehicle, performs an appropriate decoding if needed and adds additional information on data types, value types (error or measurement), and other suitable information related to the acquired data. To support easy access to individual pieces of information, the gathered and enriched data is associated to functional and hardware modules in the vehicle. In this process, a trade-off between a purely functional orientation and a hardware orientation is made. Functional structuring optimally supports the application and works well with centralized platforms. Still, the electronics in the vehicle are explicitly designed to perform distributed control and safety monitoring and to support a high degree of modularity. Thus, purely functional structuring of the information base can result in difficulties if components are updated or exchanged. As a result, the information base takes the physical separations between different hardware modules into account to structure the data. Within the hardware modules, a functional organization of information is implemented. To facilitate a purely functional approach for applications that retrieve information from the information base, several request modes to select pieces of information driven by functionality are implemented. For example, the user can access all values of a certain type that are measured by different hardware platforms. Technically, the information base acts as a server application and provides an interface via Ethernet for multiple clients. Figure 8.18 visualizes the structure of the information base that was driven by the hardware modules and indicates a content-driven request. The modules in the figure represent the top-level ECUs controlling a hardware unit, such as the axle modules or the battery charging system. These ECUs are connected to the FlexRay backbone in the vehicle.

The outlined structuring of the information base features several advantages

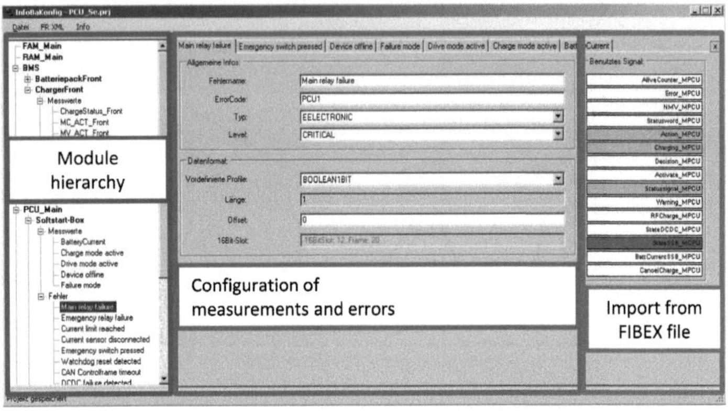

Figure 8.19.: Configuration tool for the information base

for application in the vehicle. To start with, the information base can be set up dynamically at vehicle start up. Then, each main module is loaded either from a configuration file or from the module controller via the FlexRay bus. If all controllers are loaded from configuration files, the information base does not require any transmission bandwidth on the FlexRay bus. If the controllers dynamically register during start-up, bandwidth is consumed only during that phase. In regular operation, the information base extracts all needed pieces of information from the existing bus traffic. This dynamic generation of the information base supports frequent exchange of modules. It also enables compatibility checks between the modules and supports identification of missing modules before vehicle start-up.

Configuring the information base

Accompanying the information base, a configuration tool ❸ was developed to generate the mentioned description files. The tool supports configuring the modules and is able to import base data from standardized files to describe the onboard communication[28]. As shown in Fig. 8.19, the developer can build up the logical structure of a module in a tree view and afterwards associate measurements from the data bus to these elements. Each element can then, e.g., be associated a type and a category. Then, appropriate configuration files and C-code are generated for the related controllers. The C-code is needed for online registration of the controller and can be drag-and-dropped into the controller software. Also, configurations of individual controllers can be ex- and imported to support development.

Knowledge base

The information provided by the information base is used by the knowledge base to build up the skill hierarchy for the self-concept. Figure 8.20 visualizes the hierarchical structure. At the lowest level, information is requested from the information base. These pieces of information are enriched by additional information on diag-

[28]In the MOBILE project, an importer for Field Bus Exchange Format (FIBEX)-files describing the FlexRay communication was implemented. For more details on this standard refer to the Association for Standardisation of Automation and Measuring Systems, ASAM e. V. [2008].

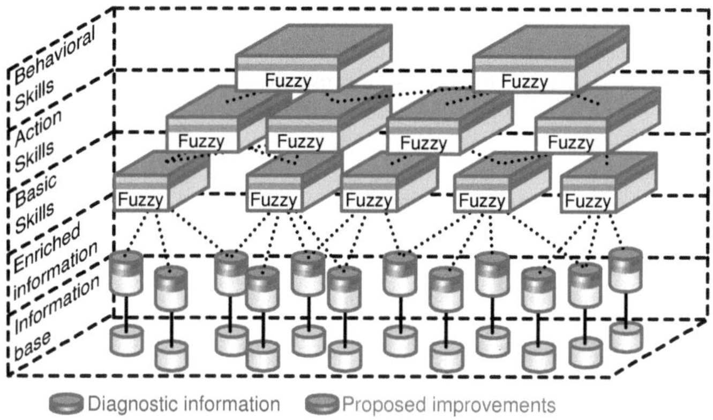

Figure 8.20.: Internal hierarchical structure of the knowledge base

nostics and proposed improvements to "tune" the vehicle. The enriched pieces of information already include a certain degree of background information on dependencies within the overall vehicle. Starting with the next higher hierarchical level, skills are introduced. At the level of basic skills, each skill is represented as a combination of enriched pieces of information and the properties of the ego vehicle. To derive the skill level, the mentioned Fuzzy Logic approach is followed. Analogously, skills are derived at higher levels. Most importantly, a skill does not necessarily require all subordinate skills to be fully available or available at all. For example, the basic steering can be achieved by all-wheel-steering, pure front-wheel steering, or rear-wheel steering. Also, the performance of a regular front-wheel steering system can be achieved by the combination of front and rear-wheel steering even if the performance of the individual steering units is limited due to design or failure. In particular, this flexibility to model soft and highly complex dependencies between skills is a major strength of the proposed approach. To support the configuration of the performance levels of a skill by the developer, a reference for the skill levels is useful. Therefore, the performance of a basic front-wheel steer, front-wheel drive mid-range vehicle with average power and weight is assumed as a basis. The reference model does not need to be modeled in detail for all characteristics, but the description should be sufficient to give the developer, who sets up the knowledge base, a sufficient understanding of important features and properties. A short description of the reference vehicle used for MOBILE can be found in A.7. The given example criteria to describe the vehicle were extracted from the automobile magazines "Auto, Motor und Sport" and "Auto Zeitung". Advanced environmental perception systems, as needed for sophisticated driver assistance systems or (partially) automated driving, are not yet considered by the reference used in the MOBILE project, but can be introduced when such skills become available. To extend the knowledge base, diagnostic information and suggestions for improvement of the vehicle can be added at any hierarchical layer. Then, if the vehicle partially fails or if the user wishes to improve the vehicle in a way he wants, he is shown possible solution strategies to repair or improve the vehicle. Due to the hierarchical organization of the knowledge base, errors and suggested improvements are inherited bottom-up, but additional information can also be added at each layer. As a result, the user can identify malfunctions at a freely selectable level of abstraction and include more or less technical details.

A fuzzy way to determine skill levels

The individual nodes in the hierarchical skill tree have to model complex dependencies among components throughout the vehicle based on expert knowledge. To model these dependencies, human thinking hugely profits from the superb abilities to abstract and generalize structures, problems, and solution strategies. In research, expert systems try to imitate these abilities for a well defined task and frequently allow to solve highly complex tasks fast, efficiently, and with sufficient accuracy [Fink and Lusth, 1987]. This work faces the challenge that the overall vehicle has to be modeled to determine the state of a high-level skill with sufficient precision. Therefore, an expert system based on Fuzzy Logic is proposed. As introduced in Sec. 7.1.1 the theory of fuzzy sets was first proposed by Zadeh [1965]. Fuzzy Logic based systems focus on the fuzzy character of the human perception and thinking as "the key elements of human thinking are not numbers, but labels of fuzzy sets, that is, classes of objects in which the transition from membership to non-membership is gradual rather than abrupt" [Zadeh, 1973, p. 1]. Consequently, the fuzzy approach focuses on the user friendliness and basic correctness of the system description or decisions rather than on the precision of the generated solution. With increasing level of abstraction of skills in the skill hierarchy, the applicability of this approach strongly increases. The core elements of a fuzzy system are fuzzy sets and appropriate membership functions. Each of these sets represents a vague concept as "high speed". The membership function associates an input value, e.g., a speed measurement, to a concept [Kruse et al., 2011, p. 255]. In general, various types of membership functions can be implemented. Still, trapezoidal functions with the special case of triangular functions are frequently used due to little computational complexity. These functions are also relied on in the MOBILE project. When both fuzzy sets and membership functions have been defined, the fuzzy evaluation process is carried out in three steps: fuzzify inputs, perform inference, and defuzzify the results to achieve sharp output values. Figure 8.21 summarizes these steps implemented in each node of the knowledge base. Details on the chosen concepts, implication, and defuzzification approaches are given in Sec. A.8. More information on Fuzzy Logic can be found in Alavala [2008], Kruse et al. [1994] or Zadeh [1965, 1973]. Two examples for application of Fuzzy Logic in a related field are introduced by Pellkofer [2003] and Schneider [2010]. Of course, the Fuzzy Logic within each node can easily be replaced by other approaches. For example classical approaches to derive decisions, e.g., based on mathematical models of the vehicle or probabilistic approaches could be used.

Evaluation

For MOBILE, a first set of demonstration skills was implemented to verify proper operation of the knowledge base. Therefore, the behavioral skills "City driving",

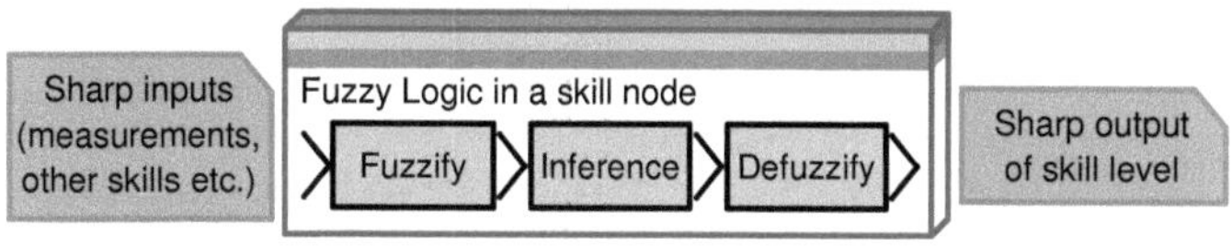

Figure 8.21.: Internal structure of a skill node with Fuzzy Logic

"Demo driving", and "Highway driving" were configured. These skills depend on several action skills and the underlying basic skills. Further skills will have to be added, as the development of MOBILE progresses. Figure 8.22 outlines the implemented skills and some relations between skills from different hierarchical layers. It is important to note that in the MOBILE project, a driver-vehicle system is considered, not yet a fully automatic system. Thus, the so far implemented skills assume that the driver is in charge of the environmental perception and vehicle guidance as the vehicle does have any means of environmental perception. This will gradually change in the future as the first sensors are being added to the vehicle in ongoing research. The dotted lines in Fig. 8.22 indicate that the strengths of the dependencies between the skills may vary with different requirements inherited from the top-level behavioral skills. For example, "City driving" requires lower speeds than "High way driving" and thus may consume less energy. Also, some dependencies may be optional and not required for the basic availability of skills. For example, maneuvering to get in or out of a parking space can significantly be improved by 4-wheel-steering, but it is not required to perform as demanded by the reference model. Up to now, the skill descriptions only regard actuators and the associated control algorithms, but more intensely than has been done in other research projects so far. Perception aspects are not yet covered, but could be added similarly to the approach presented by Siedersberger [2003] and Pellkofer [2003].

The performance levels of the behavioral skills can be the interface for the driver or a top-level autonomous system to monitor the skills of the vehicle but also to set the "driving style" – meaning how the vehicle shall behave in traffic. A more aggressive driving style will require faster actions leading, e.g., to higher energy consumption and increased requirements for the stability control. Accordingly, the strength of the individual dependencies between the skills varies, and it becomes obvious whether the vehicle can still perform all relevant maneuvers as desired.

When looking into details of the skill hierarchy, the splits among different layers are mostly easy to make. Still, within the layers sub-hierarchies can result, as already detected by Siedersberger [2003]. For example, the driving through construction sites can be associated to one or more top-level behavioral skills while being classified hierarchically higher than the so far introduced action skills. The modular structure and dependency descriptions used for the knowledge base sup-

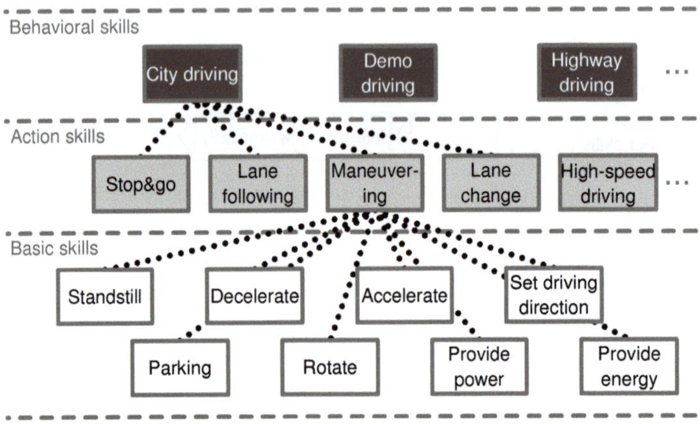

Figure 8.22.: Prototypical skill hierarchy implemented on MOBILE; the so far implemented skills describe the driver-vehicle system, not a fully automatic vehicle

port such sub-hierarchies.

Implementation

The knowledge base was implemented in C++ using objects[29], which represent each individual skill and a manager unit that connects to the information base as a client and coordinates the information exchange. The elements of the graphical user interface of the knowledge base are dynamically created at start-up depending on the number of skills on the different hierarchical levels and the current skill state. For each skill, detailed diagnostic information on failures, and suggestions for improvement are added. For quantification, the performance of the ego vehicle is compared to the one of the reference vehicle. The ego vehicle is considered as fully operational if it achieves at least this performance level. Further details on the implementation can be found in the theses of Rohde [2011], Matthaei [2010], and Günther [2011]. The knowledge base has so far not been integrated with systems for automated control and thus no evaluation results in that regard can be given. In future work, a viable starting point for integration could be usage of the skill levels provided by the knowledge base to parametrize the planned maneuvers or to select one of several available routing options.

8.2.5. Criticism and Outlook

The presented approach to self-representation features several key advantages especially for modular and flexible vehicles as MOBILE. Still, the implementation is preliminary in several aspects and features potential for future extension.

[29]Maurer [2012] pointed out that knowledge in vehicles can nicely be modeled in an object oriented manner [Maurer, 2012, p. 51].

Technical facts

The current versions of the information and knowledge base are suitable for operation in the vehicle. The system is executable in real-time with two separate timing domains: the information base is operated at low cycle times of 4-20 ms, while the knowledge base is updated only every 0.5-1 s, because its sole purpose is user information up to now. The information base is not used to reconfigure the vehicle in case of safety critical faults. This is done on lowest hardware/software level by the distributed components as introduced. Consequently, the information and knowledge base are not subject to functional safety requirements. In future, knowledge from the self-concept could parametrize the fast low-level decision making systems. The object oriented implementation based on C++ and the modular structure support extensibility. In combination with the server-client approach and the two timing levels, the system can be adapted to a wide range of applications while keeping computational and electrical resource consumption low. In terms of data bus interfacing, the current implementation is limited to FlexRay and FIBEX files to describe the bus traffic. If other connections are needed in future, appropriate drivers and importers need to be provided.

Concept: information base

The information base running in MOBILE uses offline registration to add controllers. The mechanisms for online registration were implemented but discarded due to the limited use cases in the MOBILE project and the storage consumption on the microcontroller based ECUs. The key advantage of the online registration which ensures that the configuration of the vehicle matches the installed modules has so far not come to play, because all of the modules of MOBILE are only available in one basic version. Because the knowledge base and thus the self-concept are based on the dynamically updated information base, the self-concept adapts to the timely changes of the vehicle, which reflects the gradual changes of the self-concepts of humans to some extent. Additionally, the hierarchical structuring of the information base has proven useful for the developer, because information can efficiently be entered and be kept free of unnecessary redundancy. For example, mounting positions associated to a sensor can be inherited from a containing module. In future, the information provided by the information base can easily be used by further applications. Currently, the knowledge base is the only consumer.

Concept: knowledge base

The knowledge base contributes the hugest conceptual part to the self-concept and thus the resulting self-esteem of the vehicle. The self-concept is only represented by skills and properties, which not fully reflects the self-concept as derived for humans, but it is a first step towards such systems in vehicles. Abilities are not explicitly modeled in the knowledge base, because they are hard to quantify. For humans, one way to derive an estimate of the ability level is to assess the way how fast they acquire new skills [Weineck, 2010, p. 793], which is not suitable for MOBILE or any other automated vehicle concept found during investigations for this thesis. Certainly, the set of functions currently available in MOBILE does by far not exploit the full potential of the actuators and the available computational power. Thus, the ability level of MOBILE can be assumed to be significantly higher than the current skill level captured in the first version of the knowledge base. Further development will exploit more and more abilities. The self-esteem of the vehicle is considered to some extent by adapting the skill levels depending on the current situation and comparing the own performance with a reference, which can then influence decision making. This emulates the reflection of a human on his/her own person when he/she interacts with other people. But, transfer of the already vague self-concept for humans to the technical system by the presented approach is strongly limited, and it may also be unrewarding to transfer all aspects – assumed this would be possible. For example, strong personal bias in decision making could cause unreasonable actions resulting in risky situations. In that regard, the fuzzy approach designed to capture the developer's knowledge comes along with the risk of personal influences in the system. Nevertheless, these influences can be assumed to be limited due to the technical expertise of the person in charge. Positively, the fuzzy approach seems well-suited to model dependencies of abstracted skills due to its inherent support of vagueness.

Outlook

Starting from these considerations, the knowledge base in its current implementation seems promising for relative comparison of the skills of the vehicle with other vehicles or to benchmark the probabilities of success of different possible maneuvers relying on certain skills. The evaluation of the vehicle relative to a reference model relieves the developer of the challenging task to determine the absolute performance of the vehicle relative to a "ground truth" which is hardly available. The relative evaluation suffices to judge on the improvements or the degradation of the vehicle and provides the driver with understandable evaluation results. Nevertheless, future work also needs to consider the limitations of the implemented system. The "soft" definition of all skills via the fuzzy approach performs well for high level skills that may also take the vehicle-driver interaction into account but may become questionable for low-level skills that can well be described precisely. The proposed software system allows to implement sharp skill level extraction algorithms within the individual skill nodes but was not evaluated in that regard in this work. If such quantitative measures are introduced, the skill base can be exploited to judge whether a planned maneuver is executable at all or not. Currently, the gathered knowledge only states that the maneuver would most likely be performed better than a different maneuver or at least better/worse than by a reference vehicle. Additionally, the implemented system only covers a small part of the skills relevant for driving. Further skills targeting the perception system or aspects like crash performance are not included. Most likely, skills related to environmental perception will be easy to integrate, because the concept presented in this work is closely linked to the one introduced by Siedersberger [2003] and Pellkofer [2003], who included the perception system. For environmental perception, the introduced comparative approach may be beneficial, as especially for environmental perception an absolute ground truth will be hard to define for many tasks. Relative comparisons may be easier and may suffice to choose between different possible maneuvers.

Tooling

With the tooling character of MOBILE in mind, some other extensions to the system seem useful. The information provided by the information base could facilitate a thorough consistency check of components preventing a vehicle start if modules are missing or not fully compliant. In this process, it would be helpful to automatically link the architecture description of the vehicle with the hierarchical model configured in the information base. Then, also results of the hierarchical safety analysis presented in Cha. 6 could complement the information base.

Transfer to series vehicles

If the skill base is transferred to series vehicles, the concept of skills and abilities holds true. The main goal will be to transfer the largest possible part of the abilities of the vehicle into useful skills to generate customer benefits. This issue is currently also subject to research. The DFG funded research group "Controlling Concurrent Change" at TU Braunschweig investigates how modifications to the vehicle software can be made after selling the vehicle to the customer to maximize customer benefits from the available hardware. This approach might go together well with the incentive system integrated into the knowledge base.

Conclusion

In summary, the presented skill based self-concept contributes to capture the knowledge of the vehicle about itself, which is vital for increasingly powerful actuator set-ups and the decision making processes based on this knowledge. Especially, the blurring effects of all actuators on both the lateral and the longitudinal dynamics can be reflected and to some extent be quantified in a targeted way. Still, the proposed system is just a first step towards a meaningful and fully usable self-concept. Based on experiments with the proposed system, future work will have to identify if certain parts of a human self-concept have to be modeled in more detail for vehicles or additional aspects have to be transferred.

Part V: Evaluation

The final part of this thesis evaluates the proposed approaches and mechanisms in terms of the functional safety and applicability based on MOBILE. To conclude the thesis, a brief outlook on possible future tasks will be given.

9

Functional Safety of MOBILE[1]

"Precaution is better than cure."

Edward Coke

Concluding the contributions of this thesis, this section presents a simplified analysis of the functional safety of the vehicle control function of MOBILE using the approach introduced in Cha. 6 and a hazard analysis based on ISO 26262. The procedure given in ISO 26262 is adapted to suit the evaluation of an experimental vehicle with high functional integration that is developed from scratch: the strong dependence of the hazard analysis on the evolving system architecture requires iterative re-evaluation of hazards. And, the approach proposed by ISO 26262 is modified to regard the special circumstances during operation of the experimental vehicle on a closed off test track with well known environment.

9.1. Safe State, Hazards, and ASIL Classification

To start the analysis, Tab. 9.1 provides important assumptions that were made for the operation of the vehicle on a test track. The given hazard analysis is only valid as long as the vehicle is operated under these conditions. Essentially, a trained driver with protective suit is driving the vehicle within a well defined environment and with limited speeds. The safe state[2] of the vehicle is defined as follows: Safe state

The vehicle remains controllable for the driver until the vehicle can be safely halted.

This matches the definition of Isermann et al. [2002], who derive that "for automobiles (usually), a safe state is standstill (or low speed) at a nonhazardous place" [Isermann et al., 2002, p.69], because it can be assumed that any place on the test ground is nonhazardous. Still, the definition in the MOBILE project includes timely aspects with achieving the safe state and requires to explicitly consider the handling of the vehicle during that time interval. This rough modeling of the

[1] Parts of this section have been pre-published by the author in Bergmiller [2013].

[2] ISO 26262 defines the safe state as "the operating mode of an item without an unreasonable level of risk" [ISO 26262-1, 2011, p. 14]. Risk refers to the "combination of the probability of occurrence of harm and the severity of that harm" [ISO 26262-1, 2011, p. 13].

P.J. Bergmiller, *Towards Functional Safety in Drive-by-Wire Vehicles*,
DOI 10.1007/978-3-319-17485-3_9

safe state will have to be substantiated for quantitative evaluation in this chapter. Therefore, an online evaluated vehicle model will serve as a reference as introduced in Cha. 6.

Hazard definition

Accepting the above definition of the safe state, hazards while driving MOBILE on the test track are identified. The analysis focuses on the by-Wire control and neglects assisting functions, such as the battery management or the user information system. With regard to the environment, the most critical sections of the test track are considered. The results of the analysis are given as hazards that are evaluated in terms of ASILs depending on the expected controllability, severity[3], and exposure[4]. Table 9.2 outlines two exemplary hazards that will be used as a reference in the following. It includes a case where the initial preliminary hazard analysis had to be corrected, because the integration of the individual driving functions into the vehicle control function became obvious. Such effects result, when hazards are investigated that require intervention by the driver through the sub-functions of the vehicle control system, such as the steering or braking system. With the re-evaluation finished, the highest ASIL of a functional component of a unit is assigned to the overall unit. For MOBILE, this yields an ASIL B classification, which is taken as a reference for the evaluation introduced in the following sections.

Note:

It is important to note that the comparatively low safety classification of the experimental vehicle is based on the assumptions of a skilled test driver wearing a protective suit, the well-known environment, the low speeds, and the emergency-off system. The emergency-off system features a serial redundancy structure that consists of two emergency-off switches and is kept as simple as possible. Thus, it is assumed to be always available to the driver if a failure occurs.

Remark for series vehicles

For series vehicles, the hazard analysis for the steering or braking sub-functions would obviously yield an ASIL D classification. Richter and Köhnen [2012] and Sinha [2011] confirm this result by an analysis of these functions for electric vehicles with by-Wire design. Thus, the ASIL classification of the integrated vehicle control is not higher than the one of the individual sub-functions, because ASIL D already represents the highest possible level of functional safety requirements in the automotive domain.

9.2. Hierarchical Safety Evaluation of MOBILE

Based on the hazard analysis and the introduced assumptions, this section presents the results of the functional safety evaluation of the vehicle control function using the hierarchical approach introduced in Cha. 6. As MOBILE is a primarily student driven university project, some restrictions have to be considered:

[3] In this context, the severity gives an "estimate of the extent of harm to one or more individuals that can occur in a potentially hazardous situation" [ISO 26262-1, 2011, p. 16].

[4] Exposure classifies the frequency of being in a "an operational situation that can be hazardous if coincident with [the currently analyzed, remark by the author] failure" [ISO 26262-1, 2011, p. 6].

Table 9.1.: Assumptions for operation of the experimental vehicle on the test track

Assumpt. 1	A **skilled test driver** is driving the vehicle. The driver is capable of handling critical driving situations on a high friction surface if sufficient actuators to control the vehicle are available.
Assumpt. 2	The driver wears a **protective suit**, such as used in a formula one vehicle.
Assumpt. 3	The experimental vehicle features an **emergency-off system** that is assumed to be always available and enables the driver to cut the power of the drive motors at any time.
Assumpt. 4	The test track is locked and any halting position can be considered as safe for the passengers and the surroundings.
Assumpt. 5	The test runs are only executed in **good weather** (dry road, no rain).
Assumpt. 6	The most critical sections of the test track feature **buildings or obstacles** that are located at a **minimal distance** of 6m orthogonally to the track.
Assumpt. 7	For the most critical sections of the test track, a **speed limit** of 10m/s is set that has to be obeyed by the test driver. In combination with assumptions 5 and 6, a worst case impact speed into obstacles in case of a failure of approx. 13.9m/s (50km/h) results[a].
Assumpt. 8	High speed tests are only carried out on a **wide open test site** that allows to safely stop the vehicle with deactivated drive motors even if the braking system fails.

[a]This speed was determined based on simulation experiments with the double track vehicle model introduced in Sec. 4.2 assuming different steering concepts, distances to obstacles ranging from 6m to 10m orthogonally to the track, a high friction surface, unintended acceleration of the drive motors, and a reaction time of the driver to hit the emergency off of 0.6s. This reaction time corresponds to typical reaction times of well-trained drivers, e.g., for emergency braking [McLaughlin, 2007; Mehmood and Easa, 2009]. The given maximal impact speed was determined for a scenario where the steering angles at the front and rear axle were set to 0.35rad and 0.09rad in equal directions.

- Up to 20 students were working in parallel on the parts of the vehicle. Each student worked on a specialized field at a low hierarchical level. The students are assumed to be the "experts" for a component.
- Failure rates are not available for all parts of MOBILE. For these parts typical rates were derived from the literature. Failure rates of the software

Table 9.2.: Excerpt of the hazard and risk assessment according to ISO 26262

Hazard	Severity	S[a]	Probability of Exposure	E[b]	Controllability	C[c]	ASIL
An unintended acceleration leads to a crash.	The driver's survival is uncertain, because collisions at high speed are possible.	**S3**	An operation of the vehicle at locations where an unintended acceleration can cause collisions is likely.	**E4**	A skilled test driver can simply control the vehicle by applying the emergency-off system and/or the brakes.	**C1**	**B**
A deviation from the yaw rate reference intended by the driver leads to a crash.	Light and moderate injuries are likely at low speeds ($\leq$ 50km/h) for a driver wearing a protective suite.	**S1**	An operation of the vehicle at locations where a deviation from the yaw rate reference can cause collisions is likely.	**E4**	~~At the given low speed, a skilled test driver can normally control the vehicle by braking.~~	~~**C2**~~ C3	~~**A**~~ **B**

[a]The levels S0 to S3 classify the severity of an accident. S0 denotes lowest and S3 highest severity.

[b]The levels E0 to E4 classify the exposure. E0 denotes lowest and E4 highest exposure.

[c]The levels C0 to C3 classify the controllability. C0 denotes best and C3 worst controllability.

under development are roughly estimated based on the in-field experiences. In general, the quantitative data underlying the safety analysis of MOBILE is hugely based on estimations. Thus, the quantitative results can only indicate tendencies but suffice to evaluate different design alternatives or effects of changes to the system [Schäuffele and Zurawka, 2013, p. 207].

- Although neglected by ISO 26262, aging effects are approximated by typical bath-tub curves as given in the literature [Linß, 2005, p. 359], with slightly higher failure rates of hardware components in early phases of the lifetime of the part and a significant increase towards the end.
- Only the hierarchical layers "vehicle", "system" and "subsystem" have so far been taken into account (see also highlighting in Fig. 5.2 in Cha. 5). Some components of the lower hierarchical layers are being examined in ongoing research in the MOBILE project.
- The processes followed during the development of MOBILE do not comply with the requirements imposed by ISO 26262.

The safety evaluation of MOBILE demonstrates the applicability and benefits of the hierarchical approach. Quantitative figures roughly indicate the safety level, and especially demonstrate relative changes in failure rates after modifications to components or the architecture. Working with a group of students showed that the hierarchical approach supports splitting of the complex vehicle design task into a number of smaller work packages that are easier to handle. At the same time, the system context is kept available and traceable for all developers.

9.2.1. Assumptions and Degradation Concept for MOBILE

The safe state introduced in Sec. 9.1 needs to be detailed to serve as a basis for quantitative calculations. Especially, timely aspects and the degradation concept have to be considered. The *mission time* of MOBILE is limited to 30min. After 30min, the test driver is expected to have a break and check the vehicle. The *emergency operation interval* that has to be guaranteed after the occurrence of a failure is set to 30s. This time span suffices to safely halt MOBILE on the test track even if the failure occurred while driving at the top speed of MOBILE of approx. 160km/h (44m/s). It is assumed that *only one independent fault* at a time has to be tolerated. For MOBILE, one "point in time" is defined as a 4 ms time slot. This slot length is derived from the cycle time of the FlexRay network onboard MOBILE that facilitates the synchronization of all network nodes and cyclically triggers onboard diagnostic algorithms. A similar assumption for small diagnostic time intervals is made by Sieglin [2009]. After a failure occurred, MOBILE operates in a degraded mode of operation. In this mode, the driver still has to be able to control the vehicle although possibly with degraded performance. As introduced, a virtual reference represents the minimal performance. For each failure, it was examined, whether this minimal performance can still be guaranteed. But, these investigations could not be verified in a real vehicle under varying conditions, which limits the results to working hypotheses that have to be confirmed in follow-up research. A brief summary of the related work carried out in the MOBILE project was given in Sec. 7.2.

9.2.2. Evaluation of Complexity of the Hierarchical Approach

To analyze MOBILE, eight units were defined at "system layer": (1) the front and (2) the rear axle control system consisting of the associated fault tolerant units, (3) the user input control system (also embodied by the associated fault tolerant unit), (4,5) the two power supply systems, (6,7) the emergency off systems for the drive motors at the front and at the rear and (8) the stability control system. Due to the design of MOBILE, these systems are assumed to be insusceptible to common cause failures within the scope of the research project. If cross couplings between the systems exist, the couplings are assumed to be irrelevant during the emergency operation interval of 30s. For example, low voltage buffer batteries can compensate the loss of charging power due to a failure of the high voltage

Table 9.3.: Graphically assisted failure classification at the "vehicle layer"

System	Generalized failure states: effects of the first system failure			
FAC_Sys	destabilizing	neutral	ok	
RAC_Sys	destabilizing	neutral	ok	
EOffVA_Sys	defect off	defect on	ok	
EOffRA_Sys	defect off	defect on	ok	
PSup1_Sys	all off	12V off	48V off	HV off ... ok
PSup2_Sys	all off	12V off	48V off	HV off ... ok
SC_Sys	destabilizing	off	ok	
UI_Sys	defect	braking lost	steering lost	ok

key:
FAC_Sys / RAC_Sys: front/rear axle control system;
EOffVA_Sys / EOffRA_Sys: emergency-off system for front/rear axle;
PSup1_Sys / PSup2_Sys: power supply 1/2;
SC_Sys: stability control system;
UI_Sys: user interfacing system;
vehicle operable after a failure of the system: yes, no, no failure;

system. In particular, this independence of the elements at "system level" led to the definition of the systems as given and not to the classical system partitioning into a braking, a drive, and a steering system. For each of the chosen virtual systems, 2 to 9 generalized failure states, not including the "ok"/"no failure present" state, were defined. As a result, 31 failure states have to be evaluated at "vehicle layer". The controllability has to be assessed after a given first and second system failure occurred, summing up to 702 state transitions. For the second faults, several transitions need not be considered, because they do not furthermore impact the controllability of the vehicle, and thus workload is reduced. Additionally, some transitions are identical for more than one system and have to be considered only once. Each resulting combination of generalized failure states fully describes the state of the vehicle obviating the need to take the history into account (first order Markov-Chains). Table 9.3 shows a simplified classification of MOBILE at "vehicle layer" after the first failure of each system. As mentioned, this classification at the "vehicle layer" is a challenging and not fully solved task from a scientific point of view. But, the workload to provide the information to the evaluation system is limited by the graphical environment set up in the MOBILE project.

"Subsystem layer"

The generalized failure states on "system layer" are related to approximately 60 failure states on "subsystem layer". For analysis of these dependencies, the reaction of the system to all "first faults" has to be considered. Additionally, selected "second faults" have to be taken into account. The cases for which the effect of a second fault has to be considered are identified automatically in a top down manner from "vehicle layer". In average, approx. 100 state transitions have to be evaluated at "subsystem layer" for each of the systems of MOBILE. The number of state transitions that have to be taken into account by the developer serves as an estimate for work load and complexity. If compared with "vehicle layer", the individual researcher on "system layer" has to evaluate a similar amount of relevant combinations depending on the complexity of the individual system.

Lower layers

At lower layers (component and elementary) the number of total failure states furthermore increases but again can be accommodated due to the partitioning into virtual systems and the allocation of tasks to the experts for each component. Third party components can easily be integrated at any hierarchical level. Within the MOBILE project several such components exist (steering motors, drive motors, etc.).

Link between layers

Due to the tool support, the linkage between different hierarchical layers is automatically maintained. Also, the necessary calculations are performed by the tool. As the input tables filled by the experts had continuously been updated during the development of MOBILE, the current state of the vehicle with regard to safety and the most critical components were known at any point in time. The generalized failure states and the associated documentation support the transparency and the long-term usability of the results of the safety analysis. These state descriptions also form the basis for discussions among experts working in different fields of application and at different hierarchical levels. A further extension of the tool environment to automatically link the graphical architecture description (fault trees, reliability block diagrams) or the descriptions of state transitions (Markov Chains) with the inputs of the Excel environment would be useful. Currently, these steps are carried out manually, which is acceptable for the scope of the project.

Summary

In summary, the analysis results for MOBILE can serve as a documented and tailored safety report within the scope of the research project and support continuous monitoring of the system during further development. The tailoring of the analysis by front loading knowledge on dependencies lowers the workload compared to other hierarchically structured approaches.

9.2.3. Results: Failure Rates and Diagnostic Performance

Figure 9.1 (top) illustrates the failure rates at "vehicle layer" of MOBILE for an assumed lifetime of 1140 days as parameter for the aging algorithms. Aging of components is emulated by repeated calculations with varying failure rates, which results in the curvy shape of the failure rate graphs. As introduced in Sec. 6.2.4, the software parts that have a high probability of failure were also taken into account to consider the functional redundancies – differently from the approach in ISO 26262.

Figure 9.1 (bottom) visualizes the huge advantage of considering the interactions at "system" and "vehicle layer" for the failure handling in MOBILE. For example, the curve for "system layer" considers only the cross-compensations between the different units on "subsystem layer". At higher layers, these cross-compensations are more and more due to the functional redundancies. Thus, a highly flexible vehicle as MOBILE especially profits. Analogously, an estimate of the efficiency of the diagnostic coverage is automatically derived following the hierarchical approach.

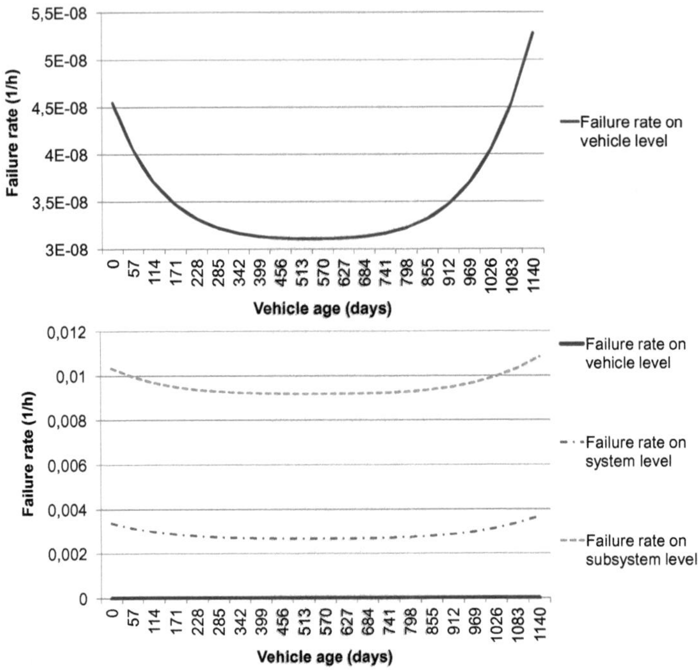

Figure 9.1.: Qualitative failure rates at "vehicle layer" (top) and "vehicle", "system" and "subsystem layer" (bottom) with increasing age of the vehicle

Tendencies in the failure rates determined in the MOBILE project show that the proposed integrated safety concept relying on functional redundancies can achieve a sufficient level of functional safety while maximizing the functional benefits from the additional actuators and limiting the system costs due to reduction in the required redundancy measures. Nevertheless, quantifying the failure rates of the control algorithms remains as a main challenge. In the MOBILE project, these failure rates are roughly set based on the literature and the experiences with the research algorithms. Thus, the expressiveness of results is limited due to the special testing conditions and the (insufficiently) small number of test runs. A similar quantification challenge will arise for environmental perception systems in automated vehicles.

10

A Step Towards Functional Safety in Drive-by-Wire Vehicles

"Don't fear failure so much that you refuse to try new things."

Louis E. Boone

Modern vehicles provide increasing functionality that more and more includes the main actuators for vehicle control. Therefore, Drive-by-Wire features tend to be integrated not only in research but also in series vehicles. This increases the complexity of the vehicle electronics both in terms of the number of devices and the complexity of the individual functions. At the same time, the functionally safe operation of the vehicle has to be assured. This thesis proposed several mechanisms which assist in closing the gap between these conflicting goals. The mechanisms are based on a top-level view of the vehicle considering it as one system rather than a collection of traditionally separated components. The contributions are application-driven and target critical aspects in terms of functional safety and functionality that have to be dealt with when implementing a Drive-by-Wire system. The processes and well-established approaches to develop functionally safe products as bundled in ISO 26262 are taken as given.

Figure 10.1 summarizes the presented contributions, which are now briefly revisited to conclude this thesis. A novel system architecture for a full Drive-by-Wire vehicle was derived in a top-down manner driven by the functional and the safety requirements by the application. The resulting architecture structures the vehicle according to these requirements rather than following the prevalent (traditional) domain-oriented approach in series vehicles. This makes it possible to exploit the full potential of the functional redundancies and the distributed monitoring to bridge the gap between the increasing functionality and functional safety.

System architecture

To enable the proposed architecture, tactical mechanisms are introduced to exploit the functional redundancies for functional safety and to perform a traceable decision making to address occurring faults. Therefore, the stability control algorithms in the literature are briefly assessed with regard to exploitation of functional redundancies among the different types of actuators. It became obvious that the quantification of the success rates of these algorithms is mandatory to enable the

Tactical mechanisms

P.J. Bergmiller, *Towards Functional Safety in Drive-by-Wire Vehicles*,
DOI 10.1007/978-3-319-17485-3_10

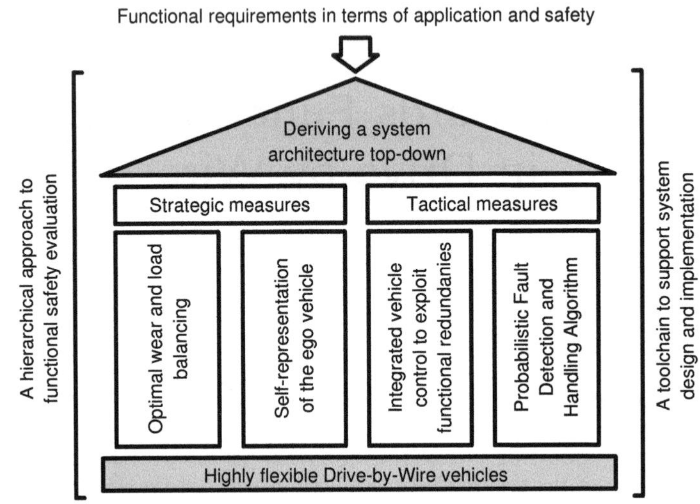

Figure 10.1.: Summary of the contributions of this thesis

integration into a functional safety evaluation. These success rates significantly impact the chances of the vehicle to achieve a safe state after a fault occurred. The safe state needs to be defined for the overall system rather than for individual components, as it is typically done. The current research is barely addressing these aspects. Still, judging from the promising results of research projects in the field of vehicle dynamics, this thesis assumes that the upcoming control algorithms will be able to exploit functional redundancies for safety purposes. To appropriately reconfigure the electronics system after a partial failure, a Probabilistic Fault Detection and Handling algorithm is introduced. The algorithm supports fast and traceable decision making. In particular, the comparative evaluation of action alternatives distinguishes the approach and supports the derivation of the most likely solution to a given problem while only requiring a low degree of redundancy. The algorithm can choose from a list of actions to treat a failure, which facilitates a well-defined degradation concept to achieve the safe state.

Strategic mechanisms

The strategic measures extend the safety concept by providing a long-term perspective on the behavior and the state of the vehicle. A system for optimal wear and load balancing aims at compensating the wear and load differences among components of different types (tires, motors, batteries) and thus reducing the probability of failure of the most loaded parts. The implemented system demonstrates the basic applicability of the approach with a focus on the easy adaptability to various applications. To capture the current state of the vehicle, a self-representation system based on the available set of skills is introduced. The system makes this knowledge available to the driver and to applications as an input for (automated)

Figure 10.2.: Research vehicle MOBILE constructed as a part of this thesis

decision making. The approach to rate skills relative to a reference vehicle or the driver's needs simplifies the evaluation of the skill level of the vehicle. Also, required upgrade or maintenance tasks are indicated. Other research projects mostly treat the ego vehicle as being always fully operational and neglect detailed knowledge about the available or degraded skills, which are vital for decision making.

Assisting tools and methods

The outlined contributions to set-up a flexible Drive-by-Wire system are accompanied by the development of a toolchain to support the application development and a hierarchical approach to analyze functional safety. The toolchain enabled the implementation of the presented algorithms both in simulation, on test benches, and in real vehicles, and is currently being re-used in other research projects. Additionally, the hierarchical approach to functional safety analysis facilitates an iterative re-evaluation of the electronics system during the development. The generalized failure states and the virtual systems facilitate tailoring of the safety analysis for complex interconnected systems and distinguish the proposed approach.

Validation and verification

All algorithms and the system structures introduced in this thesis were evaluated either in simulation, based on a 1:5 scaled vehicle, or a full-scale vehicle that was constructed as a core part of this work. As presented in the associated chapters, the strategic and tactical mechanisms to support the developed architecture could not yet be evaluated on the full-scale vehicle, but were thoroughly tested using the other given equipment. The derived architecture was successfully implemented for MOBILE (Fig. 10.2, Sec. A.9).

Summary

In summary, this thesis presents first steps to address the challenges of functional safety and complexity in modern vehicles using an automotive systems engineering approach. The introduced mechanisms represent an initial foundation and have to be further developed in future work. Nevertheless, the introduced system architecture, the tools, and the experimental vehicles with their internationally distinguished capabilities will hopefully serve as a platform for several follow-up research activities in the fields of systems engineering and vehicle control.

Erratum to: Towards Functional Safety in Drive-by-Wire Vehicles

Erratum to:
P.J. Bergmiller, *Towards Functional Safety in Drive-by-Wire Vehicles*,
DOI 10.1007/978-3-319-17485-3

The original version of this book was inadvertently published without the dissertation text on the title page.

The online version of the original book can be found under
DOI 10.1007/978-3-319-17485-3

P.J. Bergmiller (✉)
Institut für Regelungstechnik, TU Braunschweig, Braunschweig, Bayern, Germany
e-mail: bergmiller@ifr.ing.tu-bs.de

P.J. Bergmiller, *Towards Functional Safety in Drive-by-Wire Vehicles*,
DOI 10.1007/978-3-319-17485-3_11

© ... International Publishing Switzerland 20..
F. ... Ziegler, ... Food and Safety in India ...
DOI 10.1007/978-3-319-19845-8_11

A
Appendix

P.J. Bergmiller, *Towards Functional Safety in Drive-by-Wire Vehicles*,
DOI 10.1007/978-3-319-17485-3

A.1. Standards and Legislation in Germany

This section provides a brief overview of the standardization institutions, standards, and selected legislative requirements that were taken into account in this work. Contents-wise, the focus is on safety critical systems in vehicles, such as the steering and braking systems.

European Union

EWG/EG/EU directives cover various technical and non-technical topics. They are approved throughout the European Union (EU). The naming for the directives changed according to the political situation within the EU [Mittag, 2008]. EWG directives were published by the "European Economic Community" (Europäische Wirtschaftsgemeinschaft), EG and EU directives were published by the follow up institutions "European Community" (Europäische Gemeinschaft) and "European Union" (Europäische Union). The latter is eponymous since 2009.

Economic Commission for Europe

The *United Nations Economic Commission for Europe* (UNECE or ECE) is located in Genf and provides several regulations for wheeled vehicles and parts of these vehicles. Products complying to these regulations receive certification marks, which are accepted throughout the member states[1] and are acknowledged by several other nations [BMU, 2007]. The directive EU 407/2011 summarizes ECE regulations that are mandatory within the European Union. All regulations referenced in this work are included in this directive.

ISO, IEC, ITU, EN and DIN

The *International Standards Organization* (ISO) is a network of national standardization organizations [ISO, 2012], which publishes standards in all fields of application except for electrics and electronics and telecommunication. The latter two areas are covered by the *International Electrotechnical Commission* (IEC) and the *International Telecommunication Union* (ITU). European Norms (EN) and standards from the Deutsches Institut für Normung (DIN) are referenced in this work if they were not yet merged into international standards, which then replace the DIN norm.

Legislation in Germany

Within Germany, the final decision whether a vehicle may be operated on the public roads always has to be made based on the German road traffic registration ordinance ("Straßenverkehrs-Zulassungs-Ordnung"). Still, the regulations of the previously introduced organizations play a huge role in case of litigations, because they represent the state-of-the-art. Thus, any car manufacturer has to comply with these standards to avoid accusation of negligent design according to § 823 of the German Civil Code in case of system failures. Concluding, Tab. A.1 and Tab. A.2 briefly summarize the important regulations and standards published by the mentioned institutions that were referred to in this thesis.

[1] States from Europe but also Canada, the United States of America, and others

Table A.1.: Selected directives

ID	*Name*	*Description*
ECE R79	"Uniform provisions concerning the approval of vehicles with regard to steering equipment"	The document provides prescriptions for designing steering systems that enable the homologation of vehicles in one of the participating nations. This includes the definition of components of a steering system and basic technical principles to set up a steering system.
ECE R13	"Uniform provisions concerning the approval of passenger cars with regard to braking"	The document provides prescriptions for designing braking systems that enable the homologation of vehicles in one of the participating nations. This includes the definition of components of a braking system and basic technical principles to set up the system.
ECE R12	"Uniform provisions concerning the approval of vehicles with regard to the protection of the driver against the steering mechanism in the event of impact"	ECE R12 introduces requirements for the mechanical design of the steering system in order to prevent severe injuries of the driver in case of a crash.
ECE R100	"Uniform provisions concerning the approval of vehicles with regard to specific requirements for the electric power"	ECE R100 introduces regulations for the mechanical and electric design of electric vehicles with a maximal speed higher than 25km/h. Extending the technical guidelines, the labeling conventions for high voltage components are provided.

Table A.2.: Overview of relevant standards

ID	*Name*	*Description*
ISO/ IEC 2382-1	"Information technology – Vocabulary – Part1: Fundamental terms"	The standard introduces several terms, definitions, and communication conventions in the field of data and information exchange and processing.
ISO/ IEC 7498-1	"Information technology – Open Systems Interconnection - Basic Reference Model: The Basic Model"	For standardized interconnections between electronic devices, the ISO/IEC 7498-1 defines the layered "ISO/OSI"-Modell.
ISO 26262	"Road Vehicles - Functional Safety"	The standard defines requirements for the functional safety of vehicle electronics and provides guidelines for development of these systems. Especially, the processes and methods for derivation of requirements and the product development are referenced. The system structure to achieve the derived safety goals is not defined in the standard.
ISO/ IEC/ IEEE 42010	"Systems and software engineering – Architecture description"	The standard defines the nomenclature and guidelines for architecture description.
IEC 61508	"Functional Safety of Electrical/ Electronic/ Programmable Electronic Safety-related Systems"	IEC 61508 represents the basic standard for functional safety of electrics and electronics and is relevant to various fields. It references methods for risk assessment, safe product development, safe production, and documentation guidelines. Due to its general formulation, standards are derived from IEC 61508 for specific fields of application, e.g., ISO 26262 for the automotive domain.
DIN 70000	"Straßenfahrzeuge - Fahrzeugdynamik und Fahrverhalten - Begriffe"	DIN 70000 provides a common basis for naming of the measurements taken in a vehicle, the description of the vehicle handling, and the definition of coordinate systems. The standard will be succeeded by the upcoming adoption DIN ISO 8855 of ISO 8855.

A.2. The Π-Groups for Scaling of Measurements

This section introduces the Π-groups derived in this work for up- and downscaling of measurements for MAX. The Π-groups target both the lateral and the longitudinal dynamics of the vehicle. The equations (Equ. A.1) of a bicycle model (compare, e.g., [Mitschke and Wallentowitz, 2004, pp. 552]) serve as a basis to derive relevant variables and dimensions for the definition of Π-groups.

$$\begin{aligned} mv\dot{\beta} + (mv^2 + c_{\alpha_f} l_f - c_{\alpha_r} l_r)\frac{\dot{\Psi}}{v} + (c_{\alpha_f} + c_{\alpha_r})\beta &= c_{\alpha_f}\delta_f + c_{\alpha_r}\delta_r \\ \Theta\ddot{\Psi} + (c_{\alpha_f} l_f^2 + c_{\alpha_r} l_r^2)\frac{\dot{\Psi}}{v} + (c_{\alpha_f} l_f - c_{\alpha_r} l_r)\beta &= c_{\alpha_f} l_f \delta_f - c_{\alpha_r} l_r \delta_r \end{aligned} \tag{A.1}$$

Tab. A.3 summarizes the used variables and their units. As a result, three reference dimension (mass, time, length) can be found. In combination with the 13 variables given in Tab. A.3, 10 Π-groups can be defined. As repeating variables containing the reference dimensions the mass, the speed, and the distance from the front axle to the center of gravity are chosen. As a result, the Π-groups given in Equ. A.2 are

m	mass (kg)	v	speed ($\frac{m}{s}$)
Θ	inertia (kgm^2)	$\delta_{f/r}$	Steering angle front/rear (rad)
β	side slip angle (rad)	$\dot{\beta}$	side slip rate ($\frac{rad}{s}$)
$\dot{\Psi}$	yaw rate ($\frac{rad}{s}$)	$\ddot{\Psi}$	yaw acceleration ($\frac{rad}{s^2}$)
$l_{f/r}$	distance from front/rear wheel to the center of gravity (m)		
$c_{f/r}$	lateral tire stiffness front/rear ($\frac{kgm}{s^2 rad}$)		

Table A.3.: Variables and units to describe the single track model

found.

$$\begin{aligned}
\Pi_1 &= \frac{\dot{\beta} l_f}{v} \\
\Pi_2 &= \frac{\dot{\Psi} l_f}{v} \\
\Pi_3 &= \frac{\ddot{\Psi} l_f^2}{v^2} \\
\Pi_4 &= \beta \\
\Pi_5 &= \delta_f \\
\Pi_6 &= \delta_r \\
\Pi_7 &= \frac{\Theta}{l_f^2 m} \\
\Pi_8 &= \frac{c_f l}{v^2 m} \\
\Pi_9 &= \frac{c_r l}{v^2 m} \\
\Pi_{10} &= \frac{l_r}{l_f}
\end{aligned} \tag{A.2}$$

Combining the derived Π-groups and the scaling factors of MAX in terms of the mass, the length, and the speed, the multipliers given in Table 4.3 in Sec. 4.1.2 result, which have to be applied if the results obtained with the scaled vehicle need to be compared with measurements taken with a full-scale vehicle.

A.3. The Symptom-Cause Correlation Formula

The calculation of probabilities of error for the causes targeted by PFDH (Equ. 7.3) is an approximation and subject to several assumptions and restrictions, which the developer has to be aware of when applying the PFDH approach in practice.

1. To start with, it is assumed, that the survey among the members of a signal group generates a viable probability estimate of the symptom being present or not, which will be valid for most cases.

2. Additionally, the approach assumes that the conditional probabilities $P(\text{symptom}_s|\text{cause}_r)$ are high (close to one) if a dependence between the symptom and the cause is given by the cause-symptom correlation matrix. This assumption can be assumed to be valid if detailed knowledge about the hardware platform and the executed algorithms is available and considered while filling out the **CS** matrix.

3. Moreover, it is assumed that the presence of a cause is likely if the symptom pattern given in the table is detected. Thus, the conditional probability $P(\text{cause}_r| (\text{symptom}_x \cap ... \cap \text{symptom}_y))$ is assumed to be high (close to one) if the probability is calculated relative to cases where all symptoms associated to the cause in the cause-symptom correlation matrix are present. This assumption is less reliable if the cause-symptom table does not cover all causes that may trigger certain symptoms, which is likely in a practical application. If most causes are covered, the assumption becomes more reliable. The degradation approach taken in the MOBILE project based on the comparison between different alternative actions for different devices furthermore alleviates the outlined problem if it is assumed that the analysis for both redundant devices is carried out at a similar level of detail and with a similar focus in terms of the considered symptoms. As a result, the systematic miscalculation for both compared alternatives can be assumed to be similar. Still, cases where a clear symptom pattern is triggered by none of the causes listed in the matrix can potentially trigger an intervention of the system. This can be problematic if the actual (unknown) cause would require that a different action is executed. Thus, sufficient coverage of the most important faults in the cause-symptom matrix is desirable.

4. A deficit of the applied formula becomes obvious if symptom patterns can either be generated by a defined set of individual causes or a single cause generating all symptoms. For example, a cause A triggers symptom A, a cause B triggers symptom B, and a cause C triggers symptoms A and B. If the system now observes symptoms A and B it can either assume cause A and B being present or cause C. For this scenario, special measures would have to be taken, because, up to now, the system cannot distinguish between multi-point and single-point failures based on the provided information. For most

cases, a simple although theoretically not fully sound way to integrate this information would be to define the symptoms such that certain symptoms are specific to certain groups of causes if it is assumed that more than two faults will barely occur at the same time. Another approach which assumes that only single point faults can occur could include the probabilities of a symptom not being present in Equ. 7.3 if a cause is evaluated that does not trigger this symptom. Still, for practical applications this could result in a scenario where a multi-point fault cannot be associated to any cause, and thus PFDH does not trigger any actions. Theoretically, this challenge could be best addressed by extending PFDH with a combinatorial part, which checks a symptom pattern for compliance with any possible combination of causes. As this may cause significant computational effort if complex patterns are detected for huge cause-symptom correlation matrices, this approach was not followed, but the previously outlined approach following a separation of causes via symptoms was taken.

A.4. Decision Matrices for PFDH

Table A.4 outlines the cause-symptom and action-cause correlation matrices used for the first implementation of PFDH at the test bench for components of MOBILE.

Table A.4.: The cause-symptom correlation matrices for primary (left) and secondary (right) node

	Del	Prop	Limit	CANA	CANB	AliveSN	AlivePN	ADC
DevSW	1	1	1	0	0	0	0	0
IntSysTiming	1	1	0	1	1	0	0	0
CANAFail	0	0	0	1	0	0	0	0
CANBFail	0	0	0	0	1	0	0	0
FRFail	1	1	0	0	0	0	1	0
ADCFail	0	0	0	0	0	0	0	1

	Del	Prop	Limit	CANA	CANB	AliveSN	AlivePN	ADC
DelayMon	1	0	0	0	0	0	0	0
PropMon	0	1	0	0	0	0	0	0
FrequMon	0	0	0	1	1	0	0	0
LimitCheck	0	0	1	0	0	0	0	1
CANAFail	0	0	0	1	0	0	0	0
CANBFail	0	0	0	0	1	0	0	0
FRFail	1	1	0	0	0	1	1	0
InterruptUnit	0	0	0	1	1	1	0	0
ADCFail	0	0	0	0	0	0	0	1

Abbreviation	Meaning
ADC	Faulty ADC values / no values
ADCFail	Failure of ADC unit
AliveSN	Alivecounter of secondary node stopped
AlivePN	Alivecounter of primary node stopped
CANA	Missing messages on CAN channel A
CANB	Missing messages on CAN channel B
CANAFail	Failure of CAN unit A
CANBFail	Failure of CAN unit B
Del	Delay in processing of signals detected
DelayMon	Failure of delay monitoring algorithm
DevSW	Failure of software under development
FrequMon	Failure of frequency monitoring algorithm
FRFail	Failure of FlexRay unit
InterruptUnit	Failure of interrupt handler
IntSysTiming	Failure of interrupt or timing system
Limit	Signals exceed acceptable bounds
LimitCheck	Failure of monitoring algorithm to check bounds of signals
Prop	Proportional deviation in signals detected
PropMon	Failure of monitoring algorithm to detect proportional errors

Table A.5.: The action-cause correlation matrices for primary (top) and secondary (bottom) node

	DevSW	IntSysTiming	CANAFail	CANBFail	FRUnit	ADC
No action	0	0	0	0	0	0
Software reset	0.2	0.8	0.1	0.1	0.2	0
Power off	1	1	1	1	1	1

	DelayMon	PropMon	FrequMon	LimitCheck	CANAFail	CANBFail	FRFail	InterruptUnit	ADCFail
No action	0	0	0	0	0	0	0	0	0
Software reset	0.5	0.5	0.5	0.5	0.1	0.1	0.2	0.8	0
Hardware reset	0.8	0.8	0.8	0.5	0.7	0.7	0.7	0.9	0.1
Power off	1	1	1	1	1	1	1	1	1

A.5. Inductive Proof of Equivalence

Proof by Induction.

Goal:
Proof that the probability of failure p_{fail} associated to a discrete node in a Bayesian Network with a number $N \in \mathbb{N}$ of equally weighted discrete parent nodes of which the n-th parent has a failure probability of p_n can be calculated as:

$$p_{fail} = \frac{1}{N} \cdot p_1 + \frac{1}{N} \cdot p_2 + \dots + \frac{1}{N} \cdot p_n. \tag{A.3}$$

Start: $N = 1$*: (trivial)*

$$p_{fail} = \frac{1}{1} p_1 \tag{A.4}$$

Start: $N = 2$:
Assume a conditional probability table for two inputs with the weights $\frac{w_1}{w_1+w_2}$ *and* $\frac{w_2}{w_1+w_2}$ $(w_1, w_2 \in \mathbb{N})$ *for the first and second input as given in the following table:*

p_1	p_2	p_{fail}
1	1	1
1	0	$\frac{w_1}{w_1+w_2}$
0	1	$\frac{w_2}{w_1+w_2}$
0	0	0

Then, p_{fail} *is calculated as:*

$$p_{fail} = p_1 \cdot p_2 + \frac{w_1}{w_1+w_2} \cdot p_1 \cdot (1-p_2) + \frac{w_2}{w_1+w_2} \cdot (1-p_1) \cdot p_2 \tag{A.5}$$

$$= \frac{w_1}{w_1+w_2} \cdot p_1 + \frac{w_2}{w_1+w_2} \cdot p_2. \tag{A.6}$$

With all weights being equal ($w_1 = 1, w_2 = 1$ *and* $w_1 + w_2 = N = 2$),

$$p_{fail} = \frac{1}{2} \cdot p_1 + \frac{1}{2} \cdot p_2 \tag{A.7}$$

results.

Step: $n \rightarrow n+1$:
For a node with two parents and the probability

$$p_{fail} = \frac{w_1}{w_1 + w_{2_{prelimn}}} \cdot p_1 + \frac{w_{2_{prelimn}}}{w_1 + w_{2_{prelimn}}} \cdot p_{2_{prelimn}}, \tag{A.8}$$

replace the parent with the failure rate $p_{2_{prelimn}}$ *with a preliminary node with two parents, where:*

$$p_{2_{prelimn}} = \frac{w_2}{w_2+w_3} \cdot p_2 + \frac{w_3}{w_2+w_3} \cdot p_3, \tag{A.9}$$

and $w_2 + w_3 = w_{2_{prelimn}}$. *Now, the preliminary node is merged into the node of interest. The weights of the new parents are set according to the sum of weights of all inputs to the preliminary node. Starting from there, plugging Equ. A.9 in Equ. A.8 delivers the result for an incremental step in* N:

$$p_{fail} = \frac{w_1}{w_1+w_2+w_3} \cdot p_1 + \frac{w_2+w_3}{w_1+w_2+w_3} \cdot \left(\frac{w_2}{w_2+w_3} \cdot p_2 + \frac{w_3}{w_2+w_3} \cdot p_3\right) \tag{A.10}$$

$$= \frac{w_1}{w_1+w_2+w_3} \cdot p_1 + \frac{w_2}{w_1+w_2+w_3} \cdot p_2 + \frac{w_3}{w_1+w_2+w_3} \cdot p_3. \tag{A.11}$$

With all inputs to the new node being weighted equally (= 1), this iterative process to get from a node with n parents to a node with n + 1 parents can be repeated for all nodes in a Bayesian Network that have up to N inputs. As a result,

$$p_{fail} = \frac{1}{N} \cdot p_1 + \frac{1}{N} \cdot p_2 + \cdots + \frac{1}{N} \cdot p_N \tag{A.12}$$

can be used to calculate the probability of failure of a node with N inputs and a conditional probability table where each line is structured as follows:

p_1	p_2	$\cdots$	p_N	p_{fail}
$\{0,1\}$	$\{0,1\}$	$\cdots$	$\{0,1\}$	$\frac{w_{1_{new}}+w_{2_{new}}+\cdots+w_{N_{new}}}{w_1+w_2+\cdots+w_N}$

The $w_{n_{new}}$ are set to 0 if p_n is zero otherwise to w_n.

Outlook:
The above proof remains valid if individual inputs to the examined node in the Bayesian network are weighted more than others, thus $w_n \in \mathbb{R}^+$, and $\sum_{i=1}^{i=N} w_i = N$. Then, p_{fail} is calculated as follows:

$$p_{fail} = \frac{w_1}{N} \cdot p_1 + \frac{w_2}{N} \cdot p_2 + \cdots + \frac{w_N}{N} \cdot p_N. \tag{A.13}$$

q.e.d.

A.6. Proof of the Efficiency Increase

Proof by Contradiction.

Start:
For a single discrete node in a Bayesian Network with N discrete input nodes and an conditional probability table, 2^N lines have to be evaluated to derive the probability of failure of the node. For each line N-1 multiplications have to be made. The results for each of the 2^N lines have to be added up in $2^N - 1$ additions. If the formula proven in Sec. A.5 for equally weighted nodes is applied, only N multiplications and $N - 1$ additions are required.

Statement:
The number of computational operations generated to evaluate the introduced formula (Sec. A.5) is less or equal to the one generated by the evaluation of the conditional probability tables for all numbers of Nodes $N \in \mathbb{N}$. It is assumed that the number of computational operations is proportional to the computational load. Additionally, the multiplications and the additions are not treated separately.

Proof by Contradiction:
The inequality

$$N + (N-1) > (N-1) * 2^N + (2^N - 1) \tag{A.14}$$

has no solution for $N \in \mathbb{N}$.

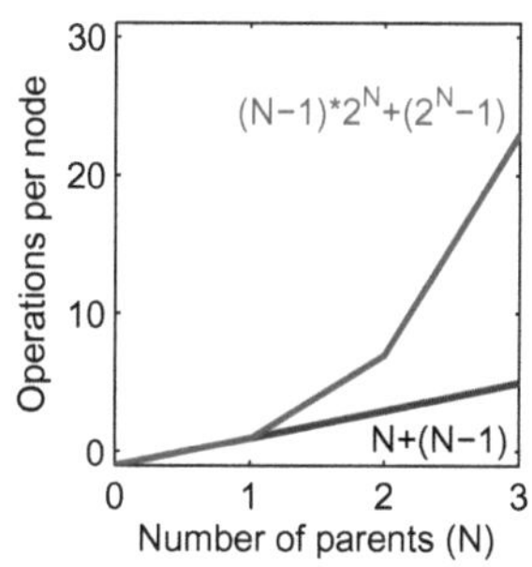

q.e.d.

A.7. The Reference Vehicle for Skill Assessment

To support the developer during evaluation of the ego-vehicle for skill assessment, a reference vehicle is introduced. The reference vehicle marks the "baseline" for an acceptable operation of the vehicle that should be maintained or surpassed during all situations. If performance below the "baseline" is detected for some maneuvers, these should be avoided. A reference vehicle could feature the following characteristics:

- front-wheel drive vehicle,
- front-wheel steering system with pre-adjusted steering ratio and a steering angle at the wheels of approx. ±40 degree,

- a track width of approx. 1.5m and a wheel base of 2.7m,
- a weight around 1.6t at a peak power of approx. 100kWs and a resulting, acceleration from 0 to 100km/h in around 10s,
- stop from 100km/h in approximately 40m,
- parking brake and hydraulic braking system with disc brakes at the front and at the rear axle,
- Anti-lock braking system and Electronic Stability Control.

Further definitions of a relevant basic equipment or characteristics of the reference vehicle can be added if they becomes necessary for definition of new skills. The above reference properties focus on the driving functionality and important characteristics directly linked to it. Based on these inputs a model of the vehicle dynamics, e.g., a bicycle model, can further detail the reference vehicle. Other properties of the vehicle, such as the size of the trunk or the crash performance, have so far been neglected. Of course, these aspects are related to important functionalities, such as the transportation of goods, and need to be covered if such skills are considered.

The reference model may vary over lifetime of the vehicle due to the general progress in technical development or the minimal quality level chosen by the driver. Still, it serves as a good basis for evaluation of the ego-vehicle. If a new reference is defined and related to the reference that has been used in the knowledge base so far, the migration effort is minimized and the migration can theoretically be done automatically. This way, also online updates of the self-esteem are possible.

A.8. Fuzzy Nodes in the Knowledge Base

As mentioned in Sec. 8.2, the knowledge base forming the self-concept for the vehicle MOBILE was implemented based on fuzzy nodes. Within each fuzzy node, three main steps are executed to fuzzify values, perform inference, and defuzzify the results to be forwarded to the next node at a hierarchically higher level. The following briefly outlines these steps as implemented for MOBILE.

Fuzzify

At first, six fuzzy sets are defined (Fig. A.1). Two categories are predefined as crisp sets[2] by the lower and upper acceptable thresholds of a value derived from the information base. The remaining concepts are easy to interpret for a human expert, and the degree of "quantization" suits the human thinking[3].

Inference

Based on the fuzzified values, a second step evaluates IF-THEN statements provided by the developer to determine the state of a skill and combines the results. In

[2] In contrast to a fuzzy set, a crisp set distinguishes only two states, e.g., 0 or 1 and no values in between [Kruse et al., 1994].

[3] According to Kruse et al. [1994], humans are able to distinguish between seven and eleven steps for a value.

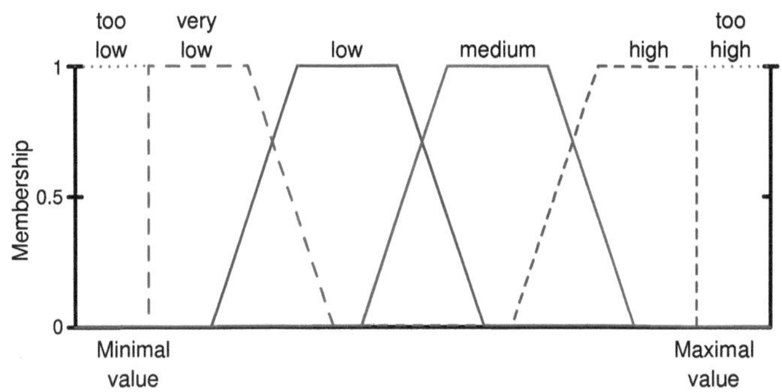

Figure A.1.: Fuzzy sets and membership functions to fuzzify input values

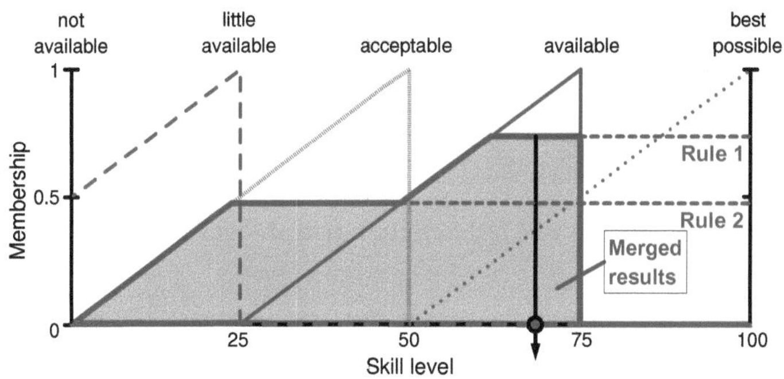

Figure A.2.: Defuzzifying the results

this, the skill level ranging from "not available" via "acceptable" to "best possible" depending on each individual IF-THEN statement is derived. Then the outputs of the individual rules are combined by using an implication rule. For the knowledge base, the Mamdani implication rule is implemented, which is essentially an "or" operation amongst the results of the individual rules [Kruse et al., 1994, p. 170].

The fuzzy skill-level output now has to be defuzzified. Therefore, the Mean-of-Maximum (MOM) approach is taken to evaluate the membership functions of the outputs, when the concept levels derived from the IF-THEN statements are applied. For MOBILE, the membership functions of the outputs were chosen such that a full association to one of the five skill levels results in a sharp skill level of 0, 25, 50, 75, or 100. 75 correlates to the reference vehicle. Figure A.2 outlines the membership functions and includes an example process to derive the sharp skill level from a given set of concept levels resulting from the previous step. It can be seen that the Mean-of-Maximum approach selects the mean value of the abscissa

Defuzzify

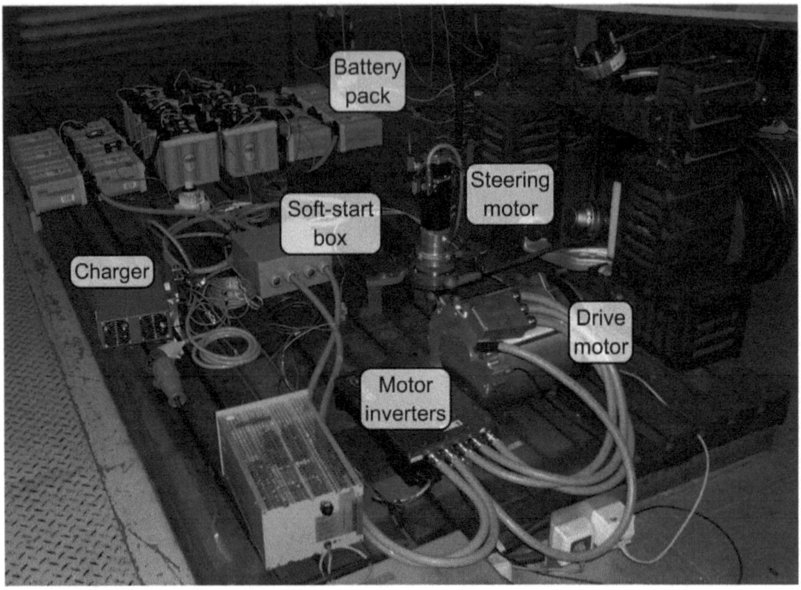

Figure A.3.: The first test bed for integration of components of MOBILE

values generating maxima in the combined output function. The resulting value quantifies the sharp skill level which is forwarded to other fuzzy nodes.

A.9. Building MOBILE

As a core part of this thesis, a full-scale experimental vehicle was designed from scratch and constructed. Starting from the outlined conceptual idea of the vehicle featuring four-wheel drive, four-wheel steering, electric brakes, and the option to integrate an active suspension system, the proposed functional architecture was developed and transferred into hardware and software components. In doing so, several steps were carried out, which included different test racks for pre-development of components and first steps towards system integration. This section briefly introduces important steps during hardware integration and vehicle construction.

High-voltage components

As a first step towards integration, starting from already pre-developed or externally sourced components, a simple test rack for the high-voltage components of MOBILE was set up (Fig. A.3). One of the two main battery packs was assembled and equipped with a balancing system. Additionally, a soft-start-box allowed to pre-charge the internal capacities of the motor inverters and implemented the emergency-off-concept. To charge the batteries, a single charger was integrated

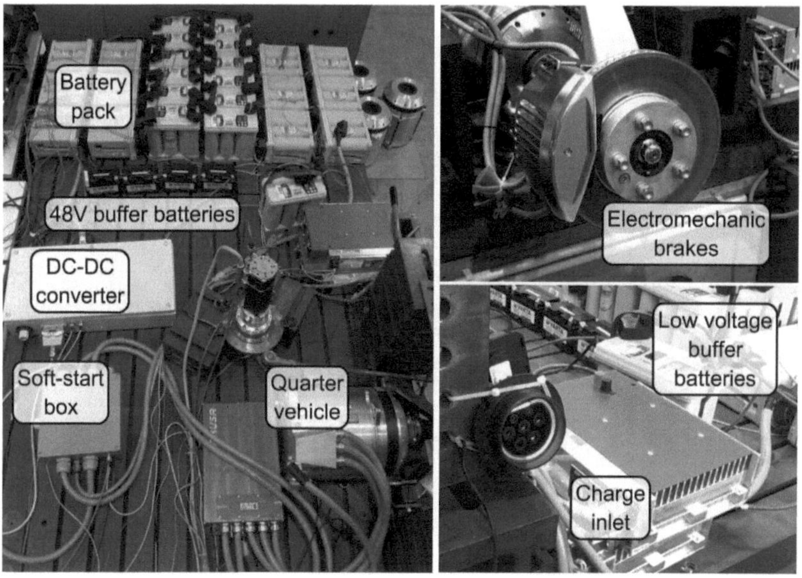

Figure A.4.: The second test bed including a "quarter vehicle" and all needed electrics and electronics

into the system. The steering system shown in Fig. A.3 was not yet integrated with the battery set-up, because all low-voltage sources were still external. Still, the set-up made it possible to drive one motor, steer a single wheel, and evaluate the performance of the high voltage components.

Electronics

In a second step (Fig. A.4), the test bench was migrated and extended by low-voltage and voltage conversion units. A DC-DC-box continuously charges the low voltage circuits at 12V and 48V from the main battery pack. Additionally, the fault tolerant units to control the front axle and the user input devices were assembled and supplemented by the mechanical/electronic user interface (gas pedal, active brake pedal, steering system, visualization). The drive and steering motors and the electromechanical braking system were integrated into a quarter-vehicle set-up. Using this test-bench, all important control functions of the vehicle were pre-developed for one power-lane and half an axle.

Vehicle frame

In parallel to the previous step, the vehicle frame was designed and evaluated in cooperation with the Institute of Engineering Design. After completion of the design, the frame was constructed at the Institute of Control Engineering as a tube frame (Fig. A.5, left).

Integration

Following, the pre-developed and tested components from the test bed given in Fig. A.4 were integrated into the mechanical frame starting with the drive motors

Figure A.5.: Frame construction and integration of high-voltage components

via the main battery packs to the low-voltage devices. Apart from the mechanical construction of the appropriate component mountings, the main focus of this step was the implementation of the second main power lane and the integration of further electronic components as the fault tolerant units for the rear axle, multiple sensors, and the second units for the DC-DC conversion and the soft starting (Fig. A.5, right). Additionally, the four suspension systems were completed and mounted to the vehicle.

Figure A.6 shows the vehicle still at the test rack. At this stage, a cooling system for the power electronics was implemented and the interior of the vehicle was completed. Thus, the basic operation of the vehicle could be verified in multiple test cases. Brakes, drive motors, and the steering system were tested during phases with heavy load (simulated load by the brakes, accelerations), and with regard to usability for the driver.

Summary

Finally, Fig. 10.2 in Cha. 10 shows the vehicle at the test site during the first test drive. In summary, the mechanical design of the vehicle starting from the construction of the vehicle frame not taking into account the test bed based evaluations took approximately 8 months with several students working on the mechanical, electric, and software integration tasks. The needed partitioning of the work packages for working in parallel on the vehicle was supported by the modular design of MOBILE, which was driven by the top-down architecture definition. At its current stage of

Figure A.6.: MOBILE jacked up at the test bench

construction not yet including an active suspension system, the vehicle contains 36 microcontroller based ECUs that interact via 8 High-Speed CAN-bus systems, several LIN-buses for battery management and a double-channel central FlexRay Backbone. For data logging and extension of the vehicle towards more complex but less time critical functions for user interaction or for automated driving, Ethernet and WLAN connections are provided via a gateway controller.

Future extensions

In future work, the vehicle can easily be extended with further components as the mentioned active suspension system, components for user interaction, sensors for environmental perception, or a hull. The low-voltage power lanes were designed to supply up to 2.4kW of continuous power and higher peak loads limited only by the low-voltage buffer batteries. This power output should suffice for several additional consumers, such as new computational platforms or sensors.

B

Own Publications and Overseen Student Research Projects

Agdas, M. (2012). *Entwurf und Implementierung eines Lade-Management-Systems für Blei-Säure-Akkumulatoren in einem Elektrofahrzeug*. bachelor thesis, Technische Universität Braunschweig.

Bagschik, G. (2011). *Entwurf und Implementierung eines Batteriemanagement-Systems zur Überwachung von Blei-Säure Traktionsbatterien in einem Elektrofahrzeug*. bachelor thesis, Technische Universität Braunschweig.

Balkan, B. (2011). *Implementierung und Test eines wahrscheinlichkeitsbasierten Fehlerbehandlungssystems für ein Drive-by-Wire Fahrzeug*. bachelor thesis, Technische Universität Braunschweig.

Bergmiller, P. (2008). *Design and Implementation of a Controller Network for a Modular By-Wire Vehicle for Safety-Critical Real-Time Applications*. Diplomarbeit, Technische Universität München.

Bergmiller, P. (2011). Generische Prototypen als Testplattform für elektronische Fahrzeugsysteme. In *auto.CITY Symposium*, Braunschweig, (presentation).

Bergmiller, P. (2013). Design and Safety Analysis of a Drive-by-Wire Vehicle. In Maurer, M. and Winner, H., editors, *Automotive Systems Engineering*, pages 147–202. Springer-Verlag, Berlin.

Bergmiller, P., Botsch, M., Speth, J., and Hofmann, U. (2008). Vehicle Rear Detection in Images with Generalized Radial -Basis-Function Classifiers. In *IEEE Intelligent Vehicles Symposium*, pages 226–233, Eindhoven, Netherlands.

Bergmiller, P., Ibele, P., Maurer, M., and Gerdes, J. C. (2011a). Development Tool for Dynamic Drive Control Systems. *ATZelektronik worldwide*, 3:1–8.

Bergmiller, P., Ibele, P., Maurer, M., and Gerdes, J. C. (2011b). Werkzeug zur Entwicklung fahrdynamischer Regelungssysteme. *ATZelektronik*, 6:60–67.

Bergmiller, P. and Maurer, M. (2011). Wahrscheinlichkeitsbasierte Fehlererkennung und -behandlung für ein Drive-by-Wire Versuchsfahrzeug. In *AUTOREG*, pages 667–678, Baden-Baden.

P.J. Bergmiller, *Towards Functional Safety in Drive-by-Wire Vehicles*,
DOI 10.1007/978-3-319-17485-3

Bergmiller, P. and Maurer, M. (2012). Flexible Versuchsträger als Testplattform für Antriebskonzepte in Elektrofahrzeugen. In Schäfer, H., editor, *Trends in der elektrischen Antriebstechnologie für Hybrid- und Elektrofahrzeuge*, pages 232–243. Expert Verlag, Renningen.

Bergmiller, P., Maurer, M., and Lichte, B. (2011c). Probabilistic Fault Detection and Handling Algorithm for Testing Stability Control Systems with a Drive-by-Wire Vehicle. In *2011 IEEE International Symposium on Intelligent Control (ISIC)*, pages 601–606, Denver (CO), USA.

Bergmiller, P., Schuldt, F., and Maurer, M. (2011d). Reifenverschleißausgleich in Elektrofahrzeugen mit funktionaler Aktorredundanz. In *13. VDI-Fachtagung Reifen-Fahrwerk-Fahrbahn*, pages 333–336, Hannover.

Bergmiller, P., Schuldt, F., and Maurer, M. (2012). Optimized Control of an Electric Vehicle With Functional Actuator Redundancy. In *2012 IEEE International Conference on Vehicular Electronics and Safety (ICVES 2012)*, pages 25–30, Istanbul, Turkey.

Bergmiller, P., Stolte, T., and Maurer, M. (2013). Hierarchische Sicherheitsbewertung von Fahrzeugen mit funktionaler Aktorredundanz. In *Funktionale Sicherheit elektrischer Antriebe in Traktionsanwendungen*, Bremen, (presentation).

Bliedung, J. (2010). *Entwurf, Implementierung und Test einer graphischen Programmierumgebung für Mikrocontroller basierend auf MATLAB/Simulink*. Studienarbeit, Technische Universität Braunschweig.

Böhme, S. (2012). *Entwicklung eines fehlertoleranten Force-Feedback-Bremspedals für ein Brake-by-Wire System*. bachelor thesis, Ostfalia Hochschule für angewandte Wissenschaften.

Freier, A. (2011). *Entwicklung einer fehlertoleranten Force-Feedback Lenkeinheit zur Erfassung von Fahrereingaben in Steer-by-Wire Systemen*. bachelor thesis, Technische Universität Braunschweig.

Gemeiner, H. (2011). *Entwicklung eines Systems zur Regelung der Fahrzeugquerdynamik basierend auf Gierrate und Schwimmwinkel*. Studienarbeit, Technische Universität Braunschweig.

Goldschmidt, Dirk (2012). *Entwicklung eines fahrdynamischen Stabilitätsprogramms für ein Drive-by-Wire Versuchsfahrzeug*. Diplomarbeit, Technische Universität Braunschweig.

Günther, T. (2011). *Entwurf und Implementierung einer Informationsbasis für intelligente Kraftfahrzeuge*. bachelor thesis, Technische Universität Braunschweig.

Güntner, A. (2012). *Entwurf und Implementierung einer frei konfigurierbaren Mensch-Maschine-Schnittstelle für dezentralen Zugriff auf Fahrzeugdaten*. Semesterarbeit, Technische Universität München.

Hackel, B. (2011). *Entwicklung einer fehlertoleranten Lenkeinheit für das Steer-by-Wire System eines vollelektrischen Versuchsfahrzeugs*. bachelor thesis, Technische Universität Braunschweig.

Hinze, M. (2012). *Entwicklung und Implementierung einer Ansteuerungslogik für aktive Redundanzmechanismen im drive-by-wire Fahrzeug auf Gesamtfahrzeugebene*. bachelor thesis, Technische Universität Braunschweig.

Homann, A. (2011). *Entwicklung einer Rahmenstruktur für die vordere Karosseriestruktur des fahrbaren Versuchsträgers MOBILE*. Studienarbeit, Technische Universität Braunschweig.

Homann, M. (2010). *Design and Implementation of a Data Acquisition Module with Protection against Electromagnetic Interference for an Experimental By-Wire Vehicle*. Diplomarbeit, Technische Universität Braunschweig.

Ibele, P. (2009). *Design and Implementation of a FlexRay-Based Interactive Dashboard for an Electric-Modular by-Wire Test-Bed-Vehicle*. Diplomarbeit, Technische Universität München.

Klingner, S. (2013). *Weiterentwicklung eines elektronischen Stabilitätsregelprogramms für ein überaktuiertes Drive-by-Wire Fahrzeug*. master thesis, Universität Braunschweig.

Laskowski, J. (2011). *Implementierung eines Data-Streaming-Konzepts für ein frei konfigurierbares Fahrzeug-Armaturenbrett*. bachelor thesis, Technische Universität Braunschweig.

Lieberam, J. (2011). *Entwicklung eines Softwaresystems zur Zustandserfassung und -regelung im Kraftfahrzeug*. Diplomarbeit, Technische Universität Braunschweig.

Matthaei, J. (2010). *Entwurf, Implementierung und Test einer dynamischen Wissensbasis für intelligente Kraftfahrzeuge*. bachelor thesis, Technische Universität Braunschweig.

Rieken, J. (2012). *Aufbau und Sicherheitsanalyse eines fehlertoleranten Bordnetzes für ein vollelektrisches Drive-by-Wire-Fahrzeug*. master thesis, Technische Universität Braunschweig.

Rohde, J. (2011). *Entwurf und Implementierung eines Fähigkeitenkonzepts für intelligente Kraftfahrzeuge*. bachelor thesis, Technische Universität Braunschweig.

Schuldt, F. (2011). *Verschleißoptimale Koordination funktional redundanter Aktorik im Elektrofahrzeug*. Diplomarbeit, Technische Universität Braunschweig.

Schwarz, S. (2012). *Torque-Vectoring im Elektrofahrzeug zur Unterstützung der Fahrzeugquerdynamik*. Studienarbeit, Technische Universität Braunschweig.

Stolte, T. (2011). *Development and Implementation of a Force Feedback System for a Steer-by-Wire Vehicle*. master thesis, Technische Universität Braunschweig.

Stolte, T., Bergmiller, P., and Maurer, M. (2014). Gewährleistung funktionaler Sicherheit durch domänenübergreifende Vernetzung von Systemen am Beispiel einer elektromechanischen Bremse. In *chassis.tech plus 2014*, pages 591–610, Munich.

Temming, C. (2011). *Implementierung eines Sicherheitskonzepts zur Realisierung aktiver Redundanz im by-wire Fahrzeug*. bachelor thesis, Technische Universität Braunschweig.

Töpler, S. (2010). *Entwicklung eines Abgleichreglers für die Fahrzeug-Längs- und Querdynamik*. Diplomarbeit, Technische Universität Braunschweig.

Volz, T. (2011). *Entwurf, Implementierung und Test einer fehlertoleranten Steuergerätesoftware zur Aktorikansteuerung im Kraftfahrzeug*. Studienarbeit, Technische Universität Braunschweig.

Wendler, J. T. (2013). *Software-in-the-Loop in a Moving Vehicle*. master thesis, Technische Universität Braunschweig.

C

Bibliography

Abe, M. (2012). Evaluation of Active Vehicle Motion Controls from Tire Energy Dissipation Points of View. In *11th International Symposium on Advanced Vehicle Control (AVEC'12)*, Seoul, Korea, (presentation).

Abe, M., Kano, Y., Suzuki, N., Hirata, J., Sugai, T., and Matsuoka, D. (2013). Tire Force Distribution Control to Reduce Energy Dissipation due to Tire Slip during Vehicle Motion for Full Drive-by-Wire Electric Vehicle. In *chassis.tech plus 2013*, pages 1–15, Munich.

Abele, A. (2012). Design and realization of an integrated safety concept based on an architecture model with the given example for the serial development of a powertrain control unit used in electric driven vehicle. In *Hybrid and Electric Vehicles*, pages 481–525, Braunschweig.

Abele, M. (2008). *Modellierung und Bewertung hochzuverlässiger Energiebordnetz-Architekturen für sicherheitsrelevante Verbraucher in Kraftfahrzeugen*. kassel university press GmbH, Kassel, (Dissertation Universität Kassel).

Adachi, M., Papadopoulos, Y., Sharvia, S., Parker, D., and Tohdo, T. (2011). An approach to optimization of fault tolerant architectures using HiP-HOPS. *Software: Practice and Experience*, 41(11):1303–1327.

Alavala, C. R. (2008). *Fuzzy Logic and Neural Networks*. New Age International (P) Limited, New Delhi, India.

Alcaraz, J. and Maroto, C. (2001). A Robust Genetic Algorithm for Resource Allocation. *Annals of Operations Research*, 102:83–109.

Anwar, S. and Niu, W. (2010). Analytical Redundancy Based Predictive Fault Tolerant Control of a Steer-By-Wire System Using Nonlinear Observer. In *2010 IEEE International Conference on Industrial Technology*, pages 477–482, Viña del Mar - Valparaiso, Chile.

Arbitmann, M., Raste, T., Lauer, P., Kelling, E., Eckert, A., and Rieth, P. E. (2011). Motion Control – Zentraler Baustein zukünftiger funktional strukturierter Domänenarchitektur im Fahrzeug. In *AUTOREG 2011*, pages 375–387, Baden-Baden.

P.J. Bergmiller, *Towards Functional Safety in Drive-by-Wire Vehicles*,
DOI 10.1007/978-3-319-17485-3

Arkin, R. C. (1989). Motor Schema-Based Mobile Robot Navigation. *The International Journal of Robotics Research*, 8(4):92–112.

Arkin, R. C. (1992). Integrating Behavioral, Perceptual and World Knowledge in Reactive Navigation. *Robotics and Autonomous Systems*, 8(4):105–122.

Arkin, R. C. and Balch, T. (1997). AuRA: Principles and Practice in Review. *Journal of Experimental and Theoretical Artificial Intelligence (JETAI)*, 9:175–189.

Armbruster, M. (2009). *Eine fahrzeugübergreifende X-by-Wire Plattform zur Ausführung umfassender Fahr- und Assistenzfunktionen*. Dr. Hut Verlag, Munich, (Dissertation Universität Stuttgart).

Armbruster, M., Zimmer, E., Lehmann, M., Reichel, R., Sieglin, E., Spiegelberg, G., and Sulzmann, A. (2006). Affordable X-By-Wire Technology Based on an Innovative Scalable E/E Platform-Concept. In *IEEE 63rd Vehicular Technology Conference*, pages 3016–3020, Melbourne, Australia.

Asano, M., Shimoyama, O., and Hashigaya, H. (1991). New Approach in Automotive Control – An Experimental Variable-Response Vehicle –. In *International Conference on Industrial Electronics, Control and Instrumentation*, pages 123–128, Kobe, Japan.

Association for Standardisation of Automation and Measuring Systems, ASAM e. V. (2008). *FIBEX – Field Bus Exchange Format, Version 3.0, release*.

AUTOSAR (2010). Technical Safety Status Report. Technical report.

AUTOSAR (2012). Release 4.0 Overview and Revision History. Technical report.

Bäck, T., Fogel, D. B., and Michalewics, Z., editors (1997). *Handbook of Evolutionary Computation*. Institute of Physics Publishing, Bristol, UK.

Balkan, B. (2011). *Implementierung und Test eines wahrscheinlichkeitsbasierten Fehlerbehandlungssystems für ein Drive-by-Wire Fahrzeug*. bachelor thesis, Technische Universität Braunschweig.

Beal, C. E. and Gerdes, J. C. (2010). Experimental Validation of a Linear Model Predictive Envelope Controller in the Presence of Vehicle Nonlinearities. In *6th IFAC Symposium Advances in Automotive Control*, Munich.

Bender, K., editor (2005). *Embedded Systems – qualitätsorientierte Entwicklung*. Springer-Verlag, Berlin.

Benington, H. D. (1983). Production of Large Computer Programs. *Annals of the History of Computing*, 5(4):350–361.

Benjamin, D. (2014). Toyota Underestimated 'Deadly' Risks, EE Times, http://www.eetimes.com, accessed: June 13th 2014.

Bergholz, P. A. (2003). *Bewegungsfertigkeiten im Sportunterricht*. Dissertation, Universität Tübingen, (published online).

Bergmiller, P. (2008). *Design and Implementation of a Controller Network for a Modular By-Wire Vehicle for Safety-Critical Real-Time Applications*. Diplomarbeit, Technische Universität München.

Bergmiller, P. (2013). Design and Safety Analysis of a Drive-by-Wire Vehicle. In Maurer, M. and Winner, H., editors, *Automotive Systems Engineering*, pages 147–202. Springer-Verlag, Berlin.

Bergmiller, P., Botsch, M., Speth, J., and Hofmann, U. (2008). Vehicle Rear Detection in Images with Generalized Radial -Basis-Function Classifiers. In *IEEE Intelligent Vehicles Symposium*, pages 226–233, Eindhoven, Netherlands.

Bergmiller, P., Ibele, P., Maurer, M., and Gerdes, J. C. (2011a). Development Tool for Dynamic Drive Control Systems. *ATZelektronik worldwide*, 2011-03:60–67.

Bergmiller, P. and Maurer, M. (2011). Wahrscheinlichkeitsbasierte Fehlererkennung und -behandlung für ein Drive-by-Wire Versuchsfahrzeug. In *AUTOREG*, pages 667–678, Baden-Baden.

Bergmiller, P. and Maurer, M. (2012). Flexible Versuchsträger als Testplattform für Antriebskonzepte in Elektrofahrzeugen. In Schäfer, H., editor, *Trends in der elektrischen Antriebstechnologie für Hybrid- und Elektrofahrzeuge*, pages 232–243. Expert Verlag, Renningen.

Bergmiller, P., Maurer, M., and Lichte, B. (2011b). Probabilistic Fault Detection and Handling Algorithm for Testing Stability Control Systems with a Drive-by-Wire Vehicle. In *2011 IEEE International Symposium on Intelligent Control (ISIC)*, pages 601–606, Denver (CO), USA.

Bergmiller, P., Schuldt, F., and Maurer, M. (2011c). Reifenverschleißausgleich in Elektrofahrzeugen mit funktionaler Aktorredundanz. In *13. VDI-Fachtagung Reifen-Fahrwerk-Fahrbahn*, pages 333–336, Hannover.

Bergmiller, P., Schuldt, F., and Maurer, M. (2012). Optimized Control of an Electric Vehicle With Functional Actuator Redundancy. In *2012 IEEE International Conference on Vehicular Electronics and Safety (ICVES 2012)*, pages 25–30, Istanbul, Turkey.

Bernard, M., Buckl, C., Döricht, V., Fehling, M., Fiege, L., von Grolmann, H., Ivandic, N., Janello, C., Klein, C., Kuhn, K.-J., Platzlaff, C., Riedl, B. C., Schätz, B., and Stanek, C. (2010). *Abschlussbericht des vom Bundesministerium für Wirtschaft und Technologie geförderten Verbundvorhabens "eCar-IKT-Systemarchitektur für Elektromobilität"*. ForTISS GmbH, Garching.

Bertacchini, A., Pavan, P., Tamagnini, L., and Fergnani, L. (2005). Control of Brushless Motor with Hybrid Redundancy for Force Feedback in Steer-by-Wire Applications. In *31st Annual Conference of IEEE Industrial Electronics Society, 2005. IECON 2005.*, pages 1407–1412, Raleigh, USA.

Bishop, C. M. (2006). *Pattern Recognition and Machine Learning.* Springer Science+Business Media, LLC, New York.

Blanc, S., Bonastre, A., and Gil, P. (2009). Dependability assessment of by-wire control systems using fault injection. *Journal of Systems Architecture*, 55(2):102–113.

Bliedung, J. (2010). *Entwurf, Implementierung und Test einer graphischen Programmierumgebung für Mikrocontroller basierend auf MATLAB/Simulink.* Studienarbeit, Technische Universität Braunschweig.

BMI (2010). *V-Modell XT Bund (version of March 5th 2010).* IT-Stab des Bundesministerium des Inneren.

BMU (2007). UN Wirtschaftskommission für Europa (ECE), http://www.bmu.de, accessed: Sep 29th 2012.

BMWi (2012). Automobilindustrie-Branchenkonjunktur, http://www.bmwi.de, accessed: Sep 21st 2012.

Bock, T. (2008). *Vehicle in the Loop - Test- und Simulationsumgebung für Fahrerassistenzsysteme.* Cuvillier Verlag, Göttingen, (Dissertation Technische Universität München).

Bodendorf, F. (2006). *Daten- und Wissensmanagement.* Springer-Verlag, Berlin, 2nd edition.

Boeck, G., Bommas-Ebert, U., Brandenburger, T., Hill, T., Huppelsberg, J., Königshoff, M., Poeggel, G., Teubner, P., Ulfig, N., Voß, R., Walter, K., and Zabel, H. (2009). *Prüfungswissen Physikum.* Georg Thieme Verlag KG, Stuttgart.

Bös, K. (2003). Motorische Leistungsfähigkeit von Kindern und Jugendlichen. In Schmidt, W., Hartmann-Tewes, I., and Brettschneider, W.-D., editors, *Erster Deutscher Kinder- und Jugendsportbericht*, pages 185–207. Hofmann Karl GmbH + Co., Schorndorf, 1st edition.

Brembeck, J., Ho, L. M., Schaub, A., Satzger, C., Tobolar, J., Bals, J., and Hirzinger, G. (2011). ROMO - The Robotic Electric Vehicle. In *22nd International Symposium on Dynamics of Vehicles on Roads and Tracks*, pages 1–6, Manchester.

Brennan, S. and Alleyne, A. (2001a). Robust Scalable Vehicle Control Via Non-Dimensional Vehicle Dynamics. *Vehicle System Dynamics*, 36(4-5):255–277.

Brennan, S. and Alleyne, A. (2001b). Using a Scale Testbed. *IEEE Control Systems*, 21(3):15–26.

Brown, J. W., MacLean, R. K., Laws, S., Gadda, C., and Gerdes, J. C. (2007). Experimental Vehicle Handling Modification through Steer-by-Wire and Differential Drive. In *2007 American Control Conference*, pages 2302–2307. Ieee.

Burckhardt, M. (1993). *Fahrwerktechnik: Radschlupfregelsysteme*. Vogel Fachbuchverlag, Würzburg.

BusinessDictionary (2012). Product Development Process, http://www.businessdicationary.com, accessed: Sep 23rd 2012.

Camci, F. and Chinnam, R. B. (2005). Dynamic Bayesian Networks for Machine Diagnostics: Hierarchical Hidden Markov Models vs. Competitive Learning. In *Joint Conference on Neural Networks*, pages 1752–1757, Montreal, Canada.

Carroll, J. B. (1993). *Human cognitive abilities – A survey of factor-analytic studies*. Cambridge University Press, Cambridge.

Carsten, O. M. J. and Nilsson, L. (2001). Safety Assessment of Driver Assistance Systems. *European Journal of Transport and Infrastructure Research*, 1(3):225–243.

Castillo, E., Gutiérrez, M. J., and Hadi, A. S. (1997). *Expert Systems and Probabilistic Network Models*. Springer-Verlag New York Inc., New York.

Chen, X., Salem, M., Das, T., and Xiaoqun, C. (2008). Real Time Software-in-the-Loop Simulation for Control Performance Validation. *SIMULATION*, 84(8/9):457–471.

Cherry, M. J. (2000). Is a Market in Human Organs Necessarily Explorative. *Public Affairs Quarterly*, 14(4):337–360.

Coello, C. A., Lamont, G. B., and Van Veldhuizen, D. A. (2007). *Evolutionary Algorithms for Solving Multi-Objective Problems*. Springer Science+Business Media, LLC, New York, USA, 2nd edition.

Collins (2013). Collins English Dictionary (online version), accessed: June 17th 2013.

Collinson, R. (1999). Fly-by-wire. *Computing and Control Engineering Journal*, 10(4):141.

Cornelsen, K., Jänsch, D., Gerson, S., Nietschke, W., Maurer, M., Canders, W. R., Schumacher, W., and Meyer, H. (2011). InDrive Simulator – Innovative Tool for Simulating and Designing Complex Drive Structures in Real Operation. In *Hybrid and Electric Vehicles*, pages 166–186, Braunschweig.

Daig, I. (2006). *Male Gender Role Dysfunction – Selbstdarstellung, Geschlechtsrollenstress und Gesundheitsrisiko bei Männern im Altersvergleich*. Dissertation, Freie Universität Berlin, (published online).

Damböck, D., Farid, M., Tönert, L., and Bengler, K. (2012). Übernahmezeiten beim hochautomatisierten Fahren. In *5. Tagung Fahrerassistenz*, pages 1–12, München.

Davis, G. and Olson, M. (1985). *Management Information Systems: Conceptual Foundations, Structure, and Development*. McGraw-Hill Inc., New York, USA, 2nd edition.

DDL (2012). X1 Experimental Vehicle, http://ddl.stanford.edu, accessed: Nov 4th 2012.

Deb, K., Pratap, A., Agarwal, S., and Meyarivan, T. (2002). A Fast and Elitist Multiobjective Genetic Algorithm: NSGA-II. *IEEE Transactions on Evolutionary Computation*, 6(2):182–197.

Dickinson, M. H., Farley, C. T., Full, R. J., Koehl, M., Kram, R., and Lehman, S. (2000). How Animals Move: An Integrative View. *Science*, 288(5463):100–106.

Dilger, E., Karrelmeyer, R., and Straube, B. (2004). Fault tolerant mechatronics [automotive applications]. In *10th IEEE International On-Line Testing Symposium*, pages 214–218, Washington, DC, USA.

DIN 70000 (1994). *Straßenfahrzeuge – Fahrzeugdynamik und Fahrverhalten – Begriffe*.

DIN ISO 8855 (2011). *Straßenfahrzeuge – Fahrzeugdynamik und Fahrverhalten – Begriffe (Draft)*.

DLR (2014). In-Flight-Simulator EC 135 FHS, http://www.dlr.de, accessed: June 13th 2014.

Dominguez-Garcia, A. D., Kassakian, J. G., and Schindall, J. E. (2004). A Backup System for Automotive Steer-by-Wire, Actuated by Selective Braking. In *35th Annual IEEE Power Electronics Specialists Conference*, pages 383–388, Aachen.

Domschke, W. (2005). *Einführung in Operations Research*. Springer-Verlag, Berlin, 6th edition.

Donges, E. (2012). Fahrerverhaltensmodelle. In Winner, H., Hakuli, S., and Wolf, G., editors, *Handbuch Fahrerassistenzsysteme*, pages 15–23. Vieweg+Teubner Verlag | Springer Fachmedien Wiesbaden GmbH, Wiesbaden, 2nd edition.

Dorey, A. D., Good, M. C., and Joubert, P. N. (1980). A Variable Free Control Characteristic Vehicle. *Vehicle System Dynamics: International Journal of Vehicle Mechanics and Mobility*, 9(1):19–44.

Düser, T. (2010). *X-in-the-Loop – ein durchgängiges Validierungsframework für die Fahrzeugentwicklung am Beispiel von Antriebsstrangfunktionen und Fahrerassistenzsystemen.* Forschungsberichte des Instituts für Produktentwicklung, Karlsruhe, (Dissertation Universität Karlsruhe).

Eberspächer, H., editor (1987). *Handlexikon der Sportwissenschaft.* Rowohlt-Taschenbuch-Verlag GmbH, Reinbek bei Hamburg.

Emami-Naeini, A., Akhter, M. M., and Rock, S. M. (1988). Effect of Model Uncertainty on Failure Detection: The Threshold Selector. *IEEE Transactions on Automatic Control*, 33(12):1106–1115.

Emery, L. (1998). Design and Construction of a Variable Dynamic Vehicle. In *Proceedings of the 16th International Technical Conference on the Enhanced Safety of Vehicles*, pages 552–561, Windsor, Canada.

Euchler, M., Bonitz, T., Mitte, D., and Geyer, M. (2010). Bewertung der Fahrsicherheit eines Elektrofahrzeugs bei stationärer Kreisfahrt. *ATZ – Automobiltechnische Zeitschrift*, 2010-03:206–213.

Fink, P. K. and Lusth, J. C. (1987). Expert Systems and Diagnostic Expertise in the Mechanical and Electrical Domains. *IEEE Transactions on Systems, Man, and Cypernetics*, 17(3):340–349.

Floridi, L. (2005). Is Information Meaningful Data? *Philosophy and Phenomenological Research*, 70(2):351–370.

Freitag, G. and Kuhn, K.-J. (2012). Hochintegrierter Antrieb: Radnabenantrieb ohne Reibbremse. In Schäfer, H., editor, *Trends in der elektrischen Antriebstechnologie für Hybrid- und Elektrofahrzeuge*, pages 73–83. Expert Verlag, Renningen.

Fuhrmann, T. (1980). Baukasten-Fahrzeug der Technischen Hochschule Aachen. *Auto Motor Sport*, 14:136.

Gadda, C. D., Laws, S. M., and Gerdes, J. C. (2007). Generating Diagnostic Residuals for Steer-by-Wire Vehicles. *IEEE Transactions on Control Systems Technology*, 15(3):529–540.

Gadda, C. D., Yih, P., and Gerdes, J. C. (2004). Incorporating a Model of Vehicle Dynamics in a Diagnostic System for Steer-by-Wire Vehicles. In *7th International Symposium on Advanced Vehicle Control*, Arnhem, Netherlands.

Gasser, T. M., Arzt, C., Ayoubi, M., Bartels, A., Bürkle, L., Eier, J., Flemisch, F., Häcker, D., Hesse, T., Huber, W., Lotz, C., Maurer, M., Ruth-Schumacher, S., Schwarz, J., and Vogt, W. (2012). Rechtsfolgen zunehmender Fahrzeugautomatisierung. Technical report, Bundesanstalt für Straßenwesen, Bergisch Gladbach.

Geman, O. (2011). A Fuzzy Expert Systems Design for Diagnosis of Parkinson' s Disease. In *3rd International Conference on E-Health and Bioengineering - EHB 2011*, pages 24–27, Iasi, Romania.

Gemeiner, H. (2011). *Entwicklung eines Systems zur Regelung der Fahrzeugquerdynamik basierend auf Gierrate und Schwimmwinkel.* Studienarbeit, Technische Universität Braunschweig.

Gerrig, R. J. and Zimbardo, P. G. (2002). *Psychology and life.* Allyn and Bacon, Boston, USA.

Gerstlauer, M. (2004). *Eignung neuer Informations- und Kommunikationstechnik zur Erhöhung der Internationalität von Forschung und Entwicklung – Möglichkeiten und Grenzen.* Dissertation, Universität Bamberg, (published online).

Gertsbakh, I. (2000). *Reliability Theory With Applications to Preventive Maintenance.* Springer-Verlag, Berlin.

Gnatz, M. (2005). *Vom Vorgehensmodell zum Projektplan.* Dissertation, Technische Universität München, (published online).

Goldschmidt, D. (2012). *Entwicklung eines fahrdynamischen Stabilitätsprogramms für ein Drive-by-Wire Versuchsfahrzeug.* Diplomarbeit, Technische Universität Braunschweig.

Graham, I., Henderson-Sellers, B., and Younessi, H. (1999). *The OPEN Process Specification.* Addison-Wesley Professional, New York, USA.

Gruber, P., Sharp, R. S., and Crocombe, A. D. (2012a). Normal and shear forces in the contact patch of a braked racing tyre . Part 1 : results from a finite-element model. *Vehicle System Dynamics: International Journal of Vehicle Mechanics and Mobility*, 50(2):323–337.

Gruber, P., Sharp, R. S., and Crocombe, A. D. (2012b). Normal and shear forces in the contact patch of a braked racing tyre . Part 2 : development of a physical tyre model. *Vehicle System Dynamics: International Journal of Vehicle Mechanics and Mobility*, 50(3):339–356.

Günther, T. (2011). *Entwurf und Implementierung einer Informationsbasis für intelligente Kraftfahrzeuge.* bachelor thesis, Technische Universität Braunschweig.

Güntner, A. (2012). *Entwurf und Implementierung einer frei konfigurierbaren Mensch-Maschine-Schnittstelle für dezentralen Zugriff auf Fahrzeugdaten.* Semesterarbeit, Technische Universität München.

Hammerschall, U. (2008). *Flexible Methodenintegration in anpassbare Vorgehensmodelle.* Dissertation, Technische Universität München, (published online).

Hasan, M. S. and Anwar, S. (2008). Sliding Mode Observer Based Predictive Fault Diagnosis of a Steer-By-Wire System. In *Proceedings of the 17th International Federation of Automatic Control World Congress*, pages 8534–8539, Seoul, Korea.

Hastie, T., Tibshirani, R., and Friedman, J. (2009). *The Elements of Statistical Learning.* Springer Science+Business Media, LLC, New York, 2nd edition.

Hayama, R., Higashi, M., Kawahara, S., Nakano, S., and Kumamoto, H. (2008). Fault Tolerant Architecture of Yaw Moment Management with Steer-by-Wire, Active Braking and Driving-Torque Distribution Integrated Control. *SAE Automotive Electronics Series*, 2008-01-01.

He, L., Zong, C., and Wang, C. (2010). A Steering-By-Wire Fault-Tolerance Control Strategy Based on Multi-dimension Gauss Hidden Markov Model. In *International Conference on Intelligent Control and Information Processing*, pages 227–230, Dalian, China.

Heißing, B. (2002). Die Simulation als Tool im Produktentstehungsprozess von Kraftfahrzeugen. In *Kongress Virtual Product Creation*, Berlin, (presentation).

Heiner, G. and Thurner, T. (1998). Time-triggered architecture for safety-related distributed real-time systems in transportation systems. In *Twenty-Eighth Annual International Symposium on Fault-Tolerant Computing*, pages 402–432, Washington, DC, USA.

Heise (2012). Nissan bringt Steer-by-Wire in Serie, http://www.heise.de, accessed: Jan 13th 2013.

Herath, I., Roberts, C., Arvanitis, T. N., and Bold, A. (2007). Satisfying Design Constraints for Automotive Safety-Critical Systems. *SAE Automotive Electronics Series*, 2007-01-14.

Hermans, F. J. J. and Zarrop, M. B. (1996). Model Based Statistical Change Detection for Automotive Applications. In *IEEE International Symposium on Computer-Aided Control System Design*, pages 105–110, Dearborn, USA.

Hermes, T. and Schultze, A. (2009). Modellbasierte Softwareentwicklung in der Praxis: Ein Statusbericht. In Bäker, B., editor, *Moderne Elektronik im Kraftfahrzeug IV*, pages 220–233. Expert Verlag, Renningen.

Hilgert, J. (2005). *Anwendung der Ähnlichkeitstheorie zur experimentellen Eigenschaftsabsicherung eines Bahnplanungsverfahrens für Fahrzeugführungssysteme*. Dissertation, Universität Duisburg-Essen, (published online).

Hiroyasu, T., Miki, M., and Watanabe, S. (1999). Distributed Genetic Algorithms with a New Sharing Approach in Multiobjective Optimization Problems. In *IEEE Congress on Evolutionary Computation*, pages 69–76, Washington, DC, USA.

Hoedt, J. and Konigorski, U. (2011). Integrated Electric Vehicle Control by Differential Parameterization. In *IEEE Conference on Decision and Control and European Control Conference*, pages 2517–2522, Orlando, USA.

Hoedt, J. and Konigorski, U. (2013). Fahrdynamikregelung fehlertoleranter X-By-Wire Antriebstopologien. In *47. Regelungstechnisches Kolloquium in Boppard*, Boppard, (presentation).

Hoffmann, D. (2010). *Data Warehouse im Rahmen der Business Intelligence*. Diplomica Verlag GmbH, Hamburg.

Holzmann, F. (2008). *Adaptive Cooperation between Driver and Assistant System*. Springer-Verlag, Berlin.

Huang, H.-B., Yu, G.-Q., and Gang, Z. (2010). Wear-Optimal Design: A Tire Wear Model and Sensitivity Analysis. In *2010 International Conference on Computer, Mechatronics, Control and Electronic Engineering*, pages 414–417.

Ibele, P. (2009). *Design and Implementation of a FlexRay-Based Interactive Dashboard for an Electric-Modular by-Wire Test-Bed-Vehicle*. Diplomarbeit, Technische Universität München.

IEC 61508 (2010). *Functional Safety of Electrical/Electronic/Programmable Electronic Safety-related Systems*. 2nd edition.

Isermann, R., editor (2006). *Fahrdynamik-Regelung*. Friedr. Vieweg & Sohn Verlag | GWV Fachverlage GmbH, Wiesbaden, 1st edition.

Isermann, R. (2008). *Mechatronische Systeme*. Springer-Verlag, Berlin, 2nd edition.

Isermann, R. and Beck, M. (2011). Modellbasierte Methoden zur Erhöhung der Verfügbarkeit und Sicherheit von Fahrwerkkomponenten. pages 679–690, Baden-Baden.

Isermann, R., Schwarz, R., and Stölzl, S. (2002). Fault-Tolerant Drive-by-Wire Systems. *IEEE Control Systems Magazine*, 22(5):64–81.

ISO (2012). About ISO, http://www.iso.org, accessed: Sep 29th 2012.

ISO 26262-1 (2011). *Road vehicles – Functional Safety – Part1: Vocabulary*.

ISO 26262-5 (2011). *Road vehicles – Functional Safety – Part5: Product development at the hardware level.*

ISO/IEC 2382-1 (1993). *Information technology – Vocabulary – Part 1: Fundamental terms.*

ISO/IEC/IEEE 42010 (2011). *Systems and software engineering – Architecture description.*

Jacobson, I., Booch, G., and Rumbaugh, J. (1999). *The Unified Software Development Process.* Addison-Wesley Longman, Amsterdam.

Jansen, P., Lehmann, J., and Heil, M. (2010). Macht Bewegung schlau? Über den Einfluss der Bewegung auf kognitive Fähigkeiten. *IM*, 4, (published online).

Javadian, A., Azad, I., and Gholami, O. (2011). A New Optimum Method for Sharing Tire Forces in Electronic Stability Control System. In *Fourth International Conference on Modeling, Simulation and Applied Optimization*, pages 1–6, Kuala Lumpur, Malaysia.

Johannessen, P. (2001). SIRIUS 2001. Technical report, Department of Computer Engineering Chalmers University of Technology, Göteborg, Sweden.

Johannessen, P., Ahlström, K., and Torin, J. (2002). Conceptual Design of Distributed by-Wire Systems. *SAE Automotive Electronics Series*, 2002-01-02.

Johannessen, P., Törner, F., and Torin, J. (2004a). Actuator Based Hazard Analysis for Safety Critical Systems. *Computer Safety, Reliability, and Security*, 3219:130–141.

Johannessen, P., Törner, F., and Torin, J. (2004b). Experiences from Model Based Development of Drive-by-Wire Control Systems. In Kleinjohann, B., Gao, G. R., Kopetz, H., Kleinjohann, L., and Rettberg, A., editors, *Design Methods and Applications for Distributed Embedded Systems*, pages 103–112. Springer Science+Business Media, Inc., Boston, USA.

Kauffman, W. M., Liddell, C. J., Smith, A., and Vandyke, R. D. (1949). An Apparatus for Varying Effective Dihedral in Flight with Application to a Study of Tolerable Dihedral on A Conventional Fighter Airplane, Advisory Committee for Aeronautics. Technical Report 948.

Kelling, N. A. and Heck, W. (2002). The BRAKE Project – Centralized Versus Distributed Redundancy for Brake-by-Wire Systems. *SAE Automotive Electronics Series*, 2002-01-02.

Kemmerling, A. (2000). Selbstbewusstsein oder Selbstrepräsentation. In Sandkühler, H. J., editor, *Strukturen von Selbstrepräsentation in Natur und Kultur*, pages 21–36. Peter Lang, Frankfurt am Main.

Khan, O. H. (2007). Fuzzy Logic based design of a Diagnostic for the T56 Turboprop Engine. In *International Conference on Emerging Technologies*, pages 194–198, Patras, Greece.

Kim, M. H., Lee, S., and Lee, K. C. (2010). Kalman Predictive Redundancy System for Fault Tolerance of Safety-Critical Systems. *IEEE Transactions on Industrial Informatics*, 6(1):46–53.

Kirrmann, H. and Großpietsch, K.-E. (2002). Fehlertolerante Steuerungs- und Regelungssysteme. *at*, 50(8):362–374.

Kirwan, B. and Ainsworth, L., editors (1992). *A Guide to Task Analysis: The Task Analysis Work Group*. CRC Press, Taylor & Francis Group, Boca Raton, USA.

Kobayashi, K., Watanabe, K., and Science, C. (1995). Estimation of Absolute Vehicle Speed using Fuzzy Logic Rule-Based Kalman Filter. In *Proceedings of the 1995 American Control Conference*, pages 3086–3090, Seattle, USA.

Koehn, P., Eckrich, M., Smakman, H., and Schaffert, A. (2006). Integrated Chassis Management: Introduction into BMW's approach to ICM. *SAE Technical Paper Series*, 1(1219).

Köhler, R. (1983). Markov-Ketten und Autokorrelation in der Sprach- und Textanalyse. In Köhler, R. and Boy, J., editors, *Glottometrika 5*, pages 134–167. Studienverlag Dr. N. Brockmeyer, Bochum.

Kondo, T. (2011). Revised GMDH-type Neural Network Using Artificial Intelligence and Its Application to Medical Image Diagnosis. In *2011 IEEE Workshop On Hybrid Intelligent Models And Applications*, pages 76–83, Paris, France.

König, L., Kretschmer, M., Neubeck, J., and Wiedemann, J. (2006). Nichtlineare Lenkregelung zur automatischen Spurführung im querdynamischen Grenzbereich. In *Steuerung und Regelung von Fahrzeugen und Motoren - AUTOREG*, pages 185–196, Wiesloch.

Koski, T. and Noble, J. M. (2009). *Bayesian Networks*. John Wiley & Sons Ltd., Chichester, UK.

Kreft, S., Gausemeier, J., Berssenbrügge, J., Lorenz, W., and Trächtler, A. (2010). Integration eines voll-aktiven X-by-wire Versuchsfahrzeugs in eine VR-basierte Simulationsumgebung. In *9. Paderborner Workshop Augmented and Virtual Reality in der Produktentstehung*, pages 159–171, Paderborn.

Krüger, J., Pruckner, A., and Knobel, C. (2010). Control Allocation for Road Vehicles - a system-independent approach for integrated vehicle dynamics control. In *Aachener Kolloquium Fahrzeug- und Motorentechnik*, pages 1–13, Aachen.

Kruse, R., Borgelt, C., Klawonn, F., Moewes, C., Ruß, G., and Steinbrecher, M. (2011). *Computational Intelligence.* Vieweg+Teubner Verlag | Springer Fachmedien Wiesbaden GmbH, Wiesbaden.

Kruse, R., Gebhardt, J., and Klawonn, F. (1994). *Foundations of Fuzzy Systems.* John Wiley & Sons, Ltd., Chichester, UK.

Kuhrmann, M., Ternité, T., and Friedrich, J. (2011). *Das V-Modell XT anpassen – Anpassung und Einführung kompakt für V-Modell XT Prozessingenieure.* Springer-Verlag, Berlin.

Kurz, A. (1994). *Lernende Steuerung eines autonomen mobilen Roboters.* VDI-Verlag, Düsseldorf, (Dissertation Technische Universität Darmstadt).

Kurz, A. (1995). ALEF: An autonomous vehicle which learns basic skills and constructs maps for navigation. *Robotics and Autonomous Systems*, 14(2-3):171–183.

Laskowski, J. (2011). *Implementierung eines Data-Streaming-Konzepts für ein frei konfigurierbares Fahrzeug-Armaturenbrett.* bachelor thesis, Technische Universität Braunschweig.

Laugier, C. and Fraichard, T. (2001). Decisional Architectures for Motion Autonomy. In Vlacic, L., Parent, M., and Harashima, F., editors, *Intelligent Vehicle Technologies*, pages 333–391. Butterworth-Heinemann, Oxford, UK.

Laugier, C., Fraichard, T., Paromtchik, I. E., and Garnier, P. (1998). Sensor-Based Control Architecture for a Car-Like Vehicle. In *Proceedings of the 1998 IEEE/RSJ International Conference on Intelligent Robots and Systems*, pages 216–222, Victoria, BC, Canada.

Laumanns, N. (2007). *Integrale Reglerstruktur zur effektiven Abstimmung von Fahrdynamiksystemen.* Forschungsgesellschaft Kraftfahrwesen mbH Aachen, Aachen, (Dissertation Rheinisch-Westfaelische Technische Hochschule Aachen).

Lee, A. Y., Marriott, A. T., and Le, N. T. (1997). Variable Dynamic Testbed Vehicle: Dynamics Analysis. In *International Congress and Exposition Detroit, Michigan*, Detroit, USA.

Legler, H., Gehrke, B., Krawczyk, O., Schasse, U., Rammer, C., Leheyda, N., and Sofka, W. (2009). *Die Bedeutung der Automobilindustrie für die deutsche Volkswirtschaft im europäischen Kontext.* (published online).

Legner, C., Pelli, D., Löhe, J., Walden, J., Fischer, T., and Stein, O. (2009). *Wandel in den Wertschöpfungsstrukturen der Automobilindustrie – Konsequenzen für Prozesse und Informationssysteme.* Oestrich-Winkel, (Whitepaper, published online).

Leppin, E. and Wittmann, B. (2010). Chamäleon auf Rädern. *Der Konstrukteur*, pages 40–41.

Li, Y., Zuo, S., Lei, L., Yang, X., and Wu, X. (2011). Analysis of impact factors of tire wear. *Journal of Vibration and Control*, 18(6):833–840.

Lieberam, J. (2011). *Entwicklung eines Softwaresystems zur Zustandserfassung und -regelung im Kraftfahrzeug*. Diplomarbeit, Technische Universität Braunschweig.

Linß, G. (2005). *Qualitätsmanagement für Ingenieure*. Carl Hanser Verlag, München, 2nd edition.

Lipka, R. P. and Brinthaupt, T. M., editors (1992). *Self-Perspectives across the Life Span*. State University of New York Press, Albany, USA.

Long, L. N., Hanford, S. D., Janrathitikarn, O., Sinsley, G. L., and Miller, J. A. (2007). A Review of Intelligent Systems Software for Autonomous Vehicles. In *2007 IEEE Symposium on Computational Intelligence in Security and Defense Applications*, pages 69–76, Honolulu, USA.

Löw, P., Pabst, R., and Petry, E. (2010). *Funktionale Sicherheit in der Praxis*. dpunkt.verlag GmbH, Heidelberg, 1st edition.

Lu, B., Wu, X., Figueroa, H., and Monti, A. (2007). A Low-Cost Real-Time Hardware-in-the-Loop Testing Approach of Power Electronics Controls. *IEEE Transactions on Industrial Electronics*, 54(2):919–931.

Luenberger, D. G. (1979). *Introduction to Dynamic Systems: Theory, Models, and Applications*. John Wiley & Sons, Inc.

Lupker, H., Cheli, F., Braghin, F., Gelosa, E., and Keckman, A. (2004). Numerical Prediction of Car Tire Wear. *Tire Science and Technology*, 32(3):164–186.

Lupker, H., Montanaro, F., Donadio, D., Gelosa, E., and Vis, M. A. (2002). Truck Tyre Wear Assessment And Prediction. In *7th International Symposium on Heavy Vehicle Weights and Dimensions*, pages 275–288, Delft, The Netherlands.

Madeira, H., Costa, D., and Vieira, M. (2000). On the Emulation of Software Faults by Software Fault Injection. In *International Conference on Dependable Systems and Networks*, pages 417–426, New York, USA.

Mahmud, N., Papadopoulos, Y., and Walker, M. (2010). A Translation of State Machines to Temporal Fault Trees. In *2010 International Conference on Dependable Systems and Networks Workshops*, pages 45–51, Chicago, USA.

Maier, M. W. and Rechtin, E. (2009). *The Art of Systems Architecting*. CRC Press Taylor & Francis Group, Boca Raton, USA, 3rd edition.

Masak, D. (2010). *Der Architekturreview*. Springer-Verlag, Berlin.

Mathworks (2013). GAMULTIOBJ, http://www.mathworks.de, accessed: Jan 6th 2013.

Matthaei, J. (2010). *Entwurf, Implementierung und Test einer dynamischen Wissensbasis für intelligente Kraftfahrzeuge.* bachelor thesis, Technische Universität Braunschweig.

Maurer, M. (2000). *Flexible Automatisierung von Straßenfahrzeugen mit Rechnersehen.* VDI-Verlag, Düsseldorf, (Dissertation Universität der Bundeswehr München).

Maurer, M. (2012). Entwurf und Test von Fahrerassistenzsystemen. In Winner, H., Hakuli, S., and Wolf, G., editors, *Handbuch Fahrerassistenzsysteme*, pages 43–54. Vieweg+Teubner Verlag | Springer Fachmedien Wiesbaden GmbH, Wiesbaden, 2nd edition.

Maurer, M. (2013). Automotive Systems Engineering – A Personal Perspective. In Maurer, M. and Winner, H., editors, *Automotive Systems Engineering*, pages 17–35. Springer-Verlag, Berlin, 2012 edition.

McKenna, K. J. (1974). A Variable Response Vehicle – Description and Applications. In *Joint Automatic Control Conference*, Austin, USA.

McLaughlin, S. B. (2007). *Analytic Assessment of Collision Avoidance Systems and Driver Dynamic Performance in Rear-End Crashes and Near-Crashes.* PhD, Virginia Polytechnic Institute and State University, USA, (published online).

Mehmood, A. and Easa, S. M. (2009). Modeling Reaction Time in Car-Following Behaviour Based on Human Factors. *International Journal of Applied Science, Engineering and Technology*, 5(14):93–101.

Mehnen, J. (2005). *Mehrkriterielle Optimierungsverfahren für produktionstechnische Prozesse.* Vulkan Verlag, Essen.

Mehrle, P., Vendeg, T., Hammer, A., and Weber, M. (2012). Schlanke Zusammenarbeit in der Produktentwicklung - Best Practice-Prinzipien aus der Automobilindustrie. *Zeitschrift für wissenschaftlichen Fabrikbetrieb*, 107(5):332–338.

Meiner, K. and Schnabel, G. (1987). *Bewegungslehre – Sportmotorik.* Volk und Wissen, Berlin, DDR.

Metzinger, T. (1999). *Subjekt und Selbstmodell.* mentis Verlag, Paderborn, 2nd edition.

Meyer, D. (2003). *Modellbasierte Mehrzieloptimierung mit Neuronalen Netzen und Evolutionsstrategien.* Dissertation, Technische Universität Ilmenau, (published online).

Meyer-Tuve, H., Pietsch, R., and Heißing, B. (2007). Experimental Handling Vehicle für Lehre und Forschung. *ATZ*, (06):560–565.

Miller, P. (2007). A Prototype Distributed Architecture for Safety Critical Automotive Systems. *SAE Automotive Electronics Series*, 2007-01-16.

Mishra, P. K. and Naik, S. M. (2005). Distributed Control System Development for FlexRay-based Systems. *SAE Automotive Electronics Series*, 2005-01-12.

Mitchell, W. C., Staniforth, A., and Scott, I. (2006). Analysis of Ackermann Steering Geometry. *SAE Technical Paper Series*, 2006-01-36.

Mitschke, M. and Wallentowitz, H. (2004). *Dynamik der Kraftfahrzeuge*. Springer-Verlag, Berlin, 4th edition.

Mittag, J. (2008). *Kleine Geschichte der Europäischen Union. Von der Europaidee bis zur Gegenwart.* Aschendorff-Verlag, Münster.

Mitzlaff, M., Lang, M., Kapitza, R., and Schröder-Preikschat, W. (2010). A Membership Service for a Distributed, Embedded System Based on a Time-Triggered FlexRay Network. In *2010 European Dependable Computing Conference*, pages 155–162, Valencia, Spain.

Mokhiamar, O. and Abe, M. (2005). Experimental verification using a driving simulator of the effect of simultaneous optimal distribution of tyre forces for active vehicle handling control. *Proceedings of the Institution of Mechanical Engineers, Part D: Journal of Automobile Engineering*, 219(2):135–149.

Mokhiamar, O. and Abe, M. (2006). How the four wheels should share forces in an optimum cooperative chassis control. *Control Engineering Practice*, 14:295–304.

Motruk, B., Diemer, J., Ernst, R., Buchty, R., and Berekovic, M. (2012). IDAMC: A Many-Core Platform with Run-Time Monitoring for Mixed-Criticality. In *International Symposium on High Assurance Systems Engineering*, pages 24–31, Omaha, USA.

Moussa, G., Radwan, E., and Hussain, K. (2012). Augmented Reality Vehicle system: Left-turn maneuver study. *Transportation Research Part C: Emerging Technologies*, 21(1):1–16.

Mruk, C. J. (2006). *Self-Esteem Research, Theory, and Practice*. Springer Publishing Company, Inc., New York.

Muenchhof, M., Beck, M., and Isermann, R. (2009). Fault-tolerant actuators and drives – Structures, fault detection principles and applications. *Annual Reviews in Control*, 33(2):136–148.

Müller, K., Steinbach, T., Korf, F., and Schmidt, T. C. (2011). A Real-time Ethernet Prototype Platform for Automotive Applications. In *2011 IEEE International Conference on Consumer Electronics - Berlin (ICCE-Berlin)*, pages 221–225, Berlin.

Müller, T. C. (2011). *Neuronale Modelle zur Offboard-Diagnostik in komplexen Fahrzeugsystemen*. Verlagshaus Monsenstein und Vannerdat OHG, Münster, (Dissertation Technische Universität Braunschweig).

Murphy, K. P. (2001). The Bayes Net Toolbox for Matlab. *Computing Science and Statistics*, pages 331–351.

Murphy, K. P. (2002). *Dynamic Bayesian Networks: Representation, Inference and Learning*. PhD, University of California, Berkeley, (published online).

Neudörfer, A. (2011). *Konstruieren sicherheitsgerechter Produkte*. Springer-Verlag, Berlin.

Nötzli, M. (1987). Porsche-Forschungsauto P.E.P. *Automobil Revue: Erste schweizerische Automobilzeitung*, 18:21.

OICA (2012). 2011 Production statistics, http://oica.net, accessed: Sept 21st 2012.

Ono, E., Hattori, Y., Aizawa, H., Kato, H., Tagawa, S., and Niwa, S. (2009). Clarification and Achievement of Theoretical Limitation in Vehicle Dynamics Integrated Management. *Journal of Environment and Engineering*, 4(1):89–100.

Pacejka, H. B. (2012). *Tire and Vehicle Dynamics*. Butterworth-Heinemann, Oxford, UK, 3rd edition.

Palin, R., Ward, D., Habli, I., and Rivett, R. (2011). ISO 26262 Safety Cases: Compliance and Assurance. In *6th IET International Conference on System Safety*, pages 1–6, Birmingham, UK.

Papadopoulos, Y., McDermid, J., Sasse, R., and Heiner, G. (2001). Analysis and synthesis of the behaviour of complex programmable electronic systems in conditions of failure. *Reliability Engineering and System Safety*, 71(3):229–247.

Park, T.-J., Han, C.-S., and Lee, S.-H. (2005). Development of the electronic control unit for the rack-actuating steer-by-wire using the hardware-in-the-loop simulation system. *Mechatronics*, 15(8):899–918.

Pearl, J. (1986). Fusion, Propagation, and Structuring in Belief Networks. *Artificial Intelligence*, 29:241–288.

Pearl, J. (1988). *Probabilistic Reasoning in Intelligent Systems*. Morgan Kaufmann Publishers, Inc., San Francisco, USA.

Pellkofer, D. M. (2003). *Verhaltensentscheidung für autonome Fahrzeuge mit Blickrichtungssteuerung.* Dissertation, Universität der Bundeswehr München, (published online).

Pfeffer, P. and Harrer, M. (2011). *Lenkungshandbuch.* Vieweg+Teubner Verlag | Springer Fachmedien Wiesbaden GmbH, Wiesbaden.

Philipps, J. (2012). Kontrolle ist gut, Misstrauen ist besser: Funktionale Sicherheit für integrierte Softwarefunktionen. In Schäfer, H., editor, *Trends in der elektrischen Antriebstechnologie für Hybrid- und Elektrofahrzeuge*, pages 129–140. Expert Verlag, Renningen.

Pimentel, J. (2003). Safety-Reliability of Distributed Embedded System Fault Tolerant Units. In *IECON'03. 29th Annual Conference of the IEEE Industrial Electronics Society*, pages 945–950, Roanoke, USA.

Piyabongkarn, D., Lew, J. Y., Rajamani, R., Grogg, J. A., and Yuan, Q. (2007). On the Use of Torque-Biasing Systems for Electronic Stability Control: Limitations and Possibilities. *IEEE Transactions on Control Systems Technology*, 15(3):581–589.

Plummer, A. R. (2006). Model-in-the-Loop-Testing. *Proceedings of the Institution of Mechanical Engineers, Part I: Journal of Systems and Control Engineering*, 220(3):183–199.

Powell, M. J. D. (1978). A Fast Algorithm For Nonlinearly Constrained Optimization Calculations. *Lecture Notes in Mathematics*, 630:144–157.

Pruckner, A., Stroph, R., and Pfeffer, P. (2012). Drive-By-Wire. In Eskandarian, A., editor, *Handbook of Intelligent Vehicles*, pages 235–282. Springer-Verlag London Limited, London, UK.

Qian, H., Xu, G., Yan, J., and Xu, Y. (2011). Vehicle Structure and Omnidirectionality for Higher Space Efficiency. In *9th World Congress on Intelligent Control and Automation*, pages 638–644, Taipei, Taiwan.

Rajamani, R. (2006). *Vehicle Dynamics and Control.* Springer Science+Business Media, Inc., New York.

Ramey, J. (2012). Nissan recalls some Infiniti Q50 sedans with steer-by-wire software glitch, http://www.autonews.com, accessed: Jan 17th 2014.

Rasmussen, J. (1983). Skills, Rules, and Knowledge; Signals, Signs, and Symbols, and Other Distinctions in Human Performance Models. *IEEE Transactions on Systems, Man, and Cybernetics*, 13(3):257–266.

Rausand, M. and Hoyland, A. (2009). *System Reliability Theory – Models, Statistical Methods and Applications.* John Wiley & Sons, Inc., Hoboken, USA.

Redmill, F. (1997). *Software Projects Evolutionary vs. Big-Bang Delivery*. John Wiley & Sons Ltd., Chichester, UK.

Rehage, D., Carl, U. B., and Vahl, A. (2005). Redundancy management of fault tolerant aircraft system architectures – reliability synthesis and analysis of degraded system states. *Aerospace Science and Technology*, 9(4):337–347.

Reichard, K. M. (2004). Integrating self-health awareness in autonomous systems. *Robotics and Autonomous Systems*, 49(1-2):105–112.

Reichel, R. and Armbruster, M. (2011). X-by-Wire Plattform – Konzept und Auslegung. *at - Automatisierungstechnik*, 59(9):583–596.

Reif, K., editor (2010). *Batterien, Bordnetze und Vernetzung*. Vieweg+Teubner Verlag | Springer Fachmedien Wiesbaden GmbH, Wiesbaden.

Reinold, P., Nachtigal, V., and Trächtler, A. (2010). An Advanced Electric Vehicle for Development and Test of New Vehicle-Dynamics Control Strategies. In *6th IFAC Symposium Advances in Automotive Control*, Munich.

Reitze, C. (2004). *Closed Loop, Entwicklungsplattform für mechatronische Fahrdynamikregelsysteme*. Universitätsverlag Karlsruhe, Karlsruhe, (Dissertation Universität Karlsruhe).

Reschka, A., Nothdurft, T., Hecker, P., Lichte, B., and Maurer, M. (2012). A Surveillance and Safety System based on Performance Criteria and Functional Degradation for an Autonomous Vehicle. In *15th International IEEE Conference on Intelligent Transportation Systems (ITSC 2012)*, pages 237–242, Anchorage, USA.

Richter, D. and Köhnen, A. (2012). Sicherheitsziele für zukünftige Elektro-Fahrzeuge: Sicherheitsarchitektur für den elektrischen Antrieb basierend auf den Anforderungen der ISO 26262. In Schäfer, H., editor, *Trends in der elektrischen Antriebstechnologie für Hybrid- und Elektrofahrzeuge*, pages 95–100. Expert Verlag, Renningen.

Rieth, P. E. (2012). Das mechatronische Fahrwerk der Zukunft. In Winner, H., Hakuli, S., and Wolf, G., editors, *Handbuch Fahrerassistenzsysteme*, pages 626–631. Vieweg+Teubner Verlag | Springer Fachmedien Wiesbaden GmbH, Wiesbaden.

Ringdorfer, M. and Horn, M. (2011). Development of a Wheel Slip Actuator Controller for Electric Vehicles using Energy Recuperation and Hydraulic Brake Control. In *2011 IEEE International Conference on Control Applications (CCA)*, pages 313–318, Denver (CO).

Robert Bosch GmbH, editor (2008). *Automotive Electrics – Automotive Electronics*. Wiley-Blackwell, Oxford, UK, 5th edition.

Rohde, J. (2011). *Entwurf und Implementierung eines Fähigkeitenkonzepts für intelligente Kraftfahrzeuge.* bachelor thesis, Technische Universität Braunschweig.

Rohe, M. (2012). Entwicklung der Gesamtfahrzeugstrategie eines E-Fahrzeugprototyps mit Torque Vectoring. In Schäfer, H., editor, *Trends in der elektrischen Antriebstechnologie für Hybrid- und Elektrofahrzeuge*, pages 101–111. Expert Verlag, Renningen.

Rook, P. (1986). Controlling software projects. *Software Engineering Journal*, 1(1):7–16.

Rosenberg, M. (1965). *Society and the adolescent self-image.* Princeton University Press, Princeton, USA.

Rosenberg, M. (1985). Self-Concept and Psychological Well-Being in Adolescence. In Leahy, R. L., editor, *The development of the self*, pages 205–246. Academic Press, New York.

Ruel, P.-H. (1987). Motivation et représentation de soi. *Revue des sciences de l'éducation*, 13(2):239–259.

Saito, K. and Nakano, R. (1988). Medical Diagnostic Expert System Based on PDP Model. In *IEEE International Conference on Neuronal Networks*, pages 255–262, San Diego, USA.

Sakurai, K., Matsubara, M., and Hoshino, M. (2008). Membership Middleware for Dependable and Cost-Effective X-by-Wire Systems. *SAE Automotive Electronics Series*, 2008-01-04:1–9.

Sangiovanni-Vincentelli, A. (2007). Quo Vadis, SLD? Reasoning About the Trends and Challenges of System Level Design. *Proceedings of the IEEE*, 95(3):467–506.

Saust, F. (2014). *Verkehrseffizientes automatisiertes Fahren im Stadtverkehr am Beispiel des Projekts "Stadtpilot".* Dissertation, Technische Universität Braunschweig, not yet submitted.

Schäuffele, J. and Zurawka, T. (2013). *Automotive Software Engineering – Grundlagen, Prozesse, Methoden und Werkzeuge.* Springer Vieweg, Wiesbaden, 5th edition.

Schittkowski, K. (1985). NLPQL: A Fortran Subroutine Solving Constrained Nonlinear Programming Problems. *Annals of Operations Research 5*, 5(1-4):485–500.

Schmitt, D. P. and Allik, J. (2005). Simultaneous Administration of the Rosenberg Self-Esteem Scale in 53 Nations: Exploring the Universal and Culture-Specific Features of Global Self-Esteem. *Journal of Personality and Social Psychology*, 89(4):623–642.

Schnabel, G., Harre, H.-D., and Krug, J., editors (2011). *Trainingslehre - Trainingswissenschaften*. Meyer & Meyer Verlag, Aachen, 2nd edition.

Schneider, J. H. (2010). *Modellierung und Erkennung von Fahrsituationen und Fahrmanövern für sicherheitsrelevante Fahrerassistenzsysteme*. Universitätsverlag der TU Chemnitz, Chemnitz, (Dissertation Technische Universität Chemnitz).

Schröder, D. (2009a). *Elektrische Antriebe – Grundlagen*. Springer-Verlag, Berlin.

Schröder, J. (2009b). *Adaptive Verhaltensentscheidung und Bahnplanung für kognitive Automobile*. Universitätsverlag Karlsruhe, Karlsruhe, (Dissertation Universität Karlsruhe).

Schroer, R. (2008). Flight control goes digital [Part Two, NASA at 50]. *IEEE Aerospace and Electronic Systems Magazine*, 23(10):23–28.

Schuldt, F. (2011). *Verschleißoptimale Koordination funktional redundanter Aktorik im Elektrofahrzeug*. Diplomarbeit, Technische Universität Braunschweig.

Schwall, M. L. (2005). *Dynamic Integration of Probabilistic Information for Diagnostics and Decisions*. Dissertation, Stanford University, (published online).

Schwall, M. L. and Gerdes, J. C. (2002). A Probabilistic Approach to Residual Processing for Vehicle Fault Detection. In *Proceedings of the 2002 American Control Conference*, pages 2552–2557. American Automatic Control Council.

Schwarz, S. (2012). *Torque-Vectoring im Elektrofahrzeug zur Unterstützung der Fahrzeugquerdynamik*. Studienarbeit, Technische Universität Braunschweig.

Sen, C. and Kar, N. C. (2009). Battery Pack Modeling for the Analysis of Battery Management System of a Hybrid Electric Vehicle. In *IEEE Vehicle Power and Propulsion Conference*, pages 207–212, Dearborn, USA.

Sharp, R. S., Casanova, D., and Symonds, P. (2010). A Mathematical Model for Driver Steering Control, with Design, Tuning and Performance Results. *Vehicle System Dynamics: International Journal of Vehicle Mechatronics and Mobility*, 33(5):289–326.

Shen, Q., Jiang, B., and Cocquempot, V. (2012). Fuzzy Logic System-Based Adaptive Fault Tolerant Control for Near Space Vehicle Attitude Dynamics with Actuator Faults. *IEEE Transactions on Fuzzy Systems*, 21(2):289–300.

Siedersberger, K.-H. (2003). *Komponenten zur automatischen Fahrzeugführung in sehenden (semi-)autonomen Fahrzeugen*. Dissertation, Universität der Bundeswehr München, (published online).

Sieglin, E. (2009). *Beitrag zur Energieversorgung eines innovativen Drive-by-wire-Fahrzeugkonzepts.* Expert Verlag, Renningen, (Dissertation Technische Universität Dresden).

Singh, S. K., Hiremath, V., Ojha, V. K., and Jadhav, N. (2012). Effects of Steering System Compliance on Steered Axle Tire Wear. *SAE Technical Paper Series*, 2012-01-19.

Sinha, P. (2011). Architectural design and reliability analysis of a fail-operational brake-by-wire system from ISO 26262 perspectives. *Reliability Engineering and System Safety*, 96(10):1349–1359.

Slob, J. J. (2008). State-of-the-Art Driving Simulators, a Literature Survey. Technical report, Eindhoven University of Technology, Eindhoven, Netherlands.

Smakman, H., Köhn, I. P., and Vieler, D. H. (2008). Integrated Chassis Management – ein Ansatz zur Strukturierung der Fahrdynamikregelsysteme. In *17. Aachener Kolloquium Fahrzeug- und Motorentechnik*, pages 1–13.

Spiegelhalter, D. J., Dawid, A. P., Steffen, L. L., and Cowell, R. G. (1993). Bayesian Analysis in Expert Systems. *Statistical Science*, 8(3):219–247.

Spitta, T. (2007). Was ist Informationswirtschaft? In *VHB Jahrestagung*, pages 1–16, Paderborn.

Starke, G. (2008). *Effektive Software-Architekturen.* Carl Hanser Verlag, Munich.

Stolte, T., Bergmiller, P., and Maurer, M. (2014). Gewährleistung funktionaler Sicherheit durch domänenübergreifende Vernetzung von Systemen am Beispiel einer elektromechanischen Bremse. In *chassis.tech plus 2014*, pages 591–610, Munich.

Sundar, M. and Plunkett, D. (2006). Brake-by-Wire, Motivation and Engineering – GM Sequel. *SAE Automotive Electronics Series*, 2006-01-31.

Tkachev, O. A. (1983). Application of Markov Chains for the Reliability Analysis of Systems With a Complex Structure. *Cybernetics and Systems Analysis*, 19(5):96–101.

Töpler, S. (2010). *Entwicklung eines Abgleichreglers für die Fahrzeug-Längs- und Querdynamik.* Diplomarbeit, Technische Universität Braunschweig.

Trächtler, A. and Niewels, F. (2006). Integrierte Querdynamikregelung mit ESP, AFS und aktiven Fahrwerksystemen. In Isermann, R., editor, *Fahrdynamik-Regelung*, pages 237–251. Friedr. Vieweg & Sohn Verlag | GWV Fachverlage GmbH, Wiesbaden.

Tsitlakidis, A., Godinjak, M., Stemmelen, L., Stolpe, R., Stroop, J., Lapko, R., Galla, T., Barthel, T., Kricke, C., Feldo, M., Bott, W., Lorenz, T., Goller, A., Fried, M., and Schramm, O. (2008). FIBEX - Field Bus Exchange Format Release Version. Technical report.

Tucci-Piergiovanni, S., Mraidha, C., Wozniak, E., Lanusse, A., and Gerard, S. (2011). A UML Model-Based Approach for Replication Assessment of AUTOSAR Safety-Critical Applications. In *IEEE 10th International Conference on Trust, Security and Privacy in Computing and Communications*, pages 1176–1187, Changsha, China.

Tusar, T. and Filipie, B. (2007). Differential Evolution Versus Genetic Algorithms in Multiobjective Optimization. In *EMO 2007, LNCS 4403*, pages 257–271. Springer-Verlag, Berlin.

V-Modell (2012). V-Modell (R) XT, version 1.4, (published online).

Verma, A. K. and Ajit, S. (2010). *Reliability and Safety Engineering*. Springer-Verlag London Limited, London, UK.

Verma, R., Vecchio, D. D., and Fathy, H. K. (2008). Development of a Scaled Vehicle With Longitudinal Dynamics of an HMMWV for an ITS Testbed. *IEEE/ASME Transactions on Mechatronics*, 13(1):46–57.

Voß, S. and Gutenschwager, K. (2001). *Informationsmanagement*. Springer-Verlag, Berlin.

Voelcker-Rehage, C. (2005). Der Zusammenhang zwischen motorischer und kognitiver Entwicklung im frühen Kindesalter – Ein Teilergebnis der MODALIS-Studie. *Deutsche Zeitschrift für Sportmedizin*, 56(10):358–363.

von Vietinghoff, A. (2008). *Nichtlineare Regelung von Kraftfahrzeugen in querdynamisch kritischen Fahrsituationen*. Universitätsverlag Karlsruhe, Karlsruhe, (Dissertation Universität Karlsruhe).

Walczak, S. (2005). Artificial Neural Network Medical Decision Support Tool: Predicting Transfusion Requirements of ER Patients. *IEEE Transactions on Information Technology in Biomedicine*, 9(3):468–474.

Walker, M. and Papadopoulos, Y. (2009). Qualitative temporal analysis: Towards a full implementation of the Fault Tree Handbook. *Control Engineering Practice*, 17(10):1115–1125.

Waraus, D. (2009). Steer-by-wire system based on FlexRay protocol. In *Applied Electronics*, pages 269–272, Pilsen, Czech Republic.

Ward, A. C. and Sobek, D. K. (2014). *Lean Product and Process Development*. Lean Enterprise Institute, Cambridge, MA, USA, 2nd edition.

Wehmeyer, S., editor (2005). *Oxford Advanced Learners Dictionary of Current English.* Oxford University Press, Oxford, UK, 7th edition.

Weineck, J. (2010). *Optimales Training.* Spitta Verlag GmbH & Co. KG, Balingen, 16th edition.

Wieczorrek, H. W. and Mertens, P. (2011). *Management von IT-Projekten.* Springer-Verlag, Berlin.

Willke, H. (2004). *Einführung in das systemische Wissensmanagement.* Carl-Auer-Verlag, Heidelberg.

Wilwert, C., Navet, N., Song, Y. Q., and Simonot-Lion, F. (2005). Design of automotive X-by-Wire systems. In Zurawski, R., editor, *The Industrial Communication Technology Handbook*, pages (29–1) – (29–34). CRC Press, Taylor & Francis Group, Boca Raton, USA.

Winner, H. and Heuss, O. (2005). X-by-Wire Betätigungselemente – Überblick und Ausblick. In *Darmstädter Kolloquium Mensch und Fahrzeug*, pages 1–34, Darmstadt.

X-by-Wire Project (1998). *Brite-EuRam 111 Program. X-By-Wire – safety related fault tolerant systems in vehicles, final report.*

Yang, L., He, H., Sun, F., Shi, S., Li, Y., and Liu, L. (2010). Research of Fuzzy Logic Control Strategy for Engine Start/stop in Dual-clutch Hybrid Electric Vehicle. In *2010 Seventh International Conference on Fuzzy Systems and Knowledge Discovery*, pages 912–917, Yantai, China.

Yarin, L. P. (2012). *The Pi-Theorem – Applications to Fluid Mechanics and Heat and Mass Transfer.* Springer-Verlag, Berlin.

Yu, H., Liang, W., Kuang, M., and McGee, R. (2009). Vehicle Handling Assistant Control System via Independent Rear Axle Torque Biasing. In *Proceedings of the 2009 American Control Conference*, pages 695–700.

Zack, M. H. (1999). Managing Codified Knowledge. *Sloan Management Review*, 40(4):45–58.

Zadeh, L. A. (1965). Fuzzy Sets. *Information and Control*, 8(3):338–353.

Zadeh, L. A. (1973). Outline of a New Approach to the Analysis of Complex Systems and Decision Processes. *IEEE Transactions on Systems, Man, and Cybernetics*, SMC-3(1):28–44.

Zhang, H., Xu, H., and Peng, W. (2008). A Genetic Algorithm for Solving RCPSP. In *2008 International Symposium on Computer Science and Computational Technology*, pages 246–249, Shanghai, China.

Zhen, B., Altemare, C., and Anwar, S. (2005). Fault tolerant Steer-By-Wire road wheel control system. In *Proceedings of the 2005 American Control Conference*, pages 1619–1624, Portland, USA.

Zheng, G. and Zhao, J. (2009). Implementation of A Novel Digitally-controlled Lead-acid Battery Management System. In *2009 IEEE International Conference on Intelligent Computing and Intelligent Systems*, pages 813–817, Shanghai, China.

Zhong, G., Wen, J., Li, X., Hu, J., and Yang, M. (2010). Pioneering Low Cost Solution for Power Amplifiers Managing Car Engine Start-stop Feature. In *2010 10th IEEE International Conference on Solid-State and Integrated Circuit Technology*, pages 141–143, Shanghai, China.

Zoelch, U., Singer, M., and Hohmann, D. (2006). Entwicklungsmethodik bei der Aktivlenkung. In *Steuerung und Regelung von Fahrzeugen und Motoren – AUTOREG*, pages 727–736, Wiesloch.

Zomotor, A. (1991). *Fahrwerktechnik: Fahrverhalten*. Vogel Verlag und Druck KG, Würzburg, 2nd edition.

Zuo, G., Kumamoto, H., Nishihara, O., Hayama, R., and Nakano, S. (2005). Quantitative reliability analysis of different design alternatives for steer-by-wire system. *Reliability Engineering and System Safety*, 89(3):241–247.

MIX
Papier aus verantwortungsvollen Quellen
Paper from responsible sources
FSC® C105338

If you have any concerns about our products,
you can contact us on
ProductSafety@springernature.com

In case Publisher is established outside the EU,
the EU authorized representative is:
Springer Nature Customer Service Center GmbH
Europaplatz 3, 69115 Heidelberg, Germany

Printed by Libri Plureos GmbH
in Hamburg, Germany